ELECTROLUMINESCENCE

ELEKTROLYUMINESTSENTSIYA

ЭЛЕКТРОЛЮМИНЕСЦЕНЦИЯ

The Lebedev Physics Institute Series

Editor: Academician D. V. Skobel'tsyn

Director, P. N. Lebedev Physics Institute, Academy of Sciences of the USSR

Proceedings (Trudy) of the P. N. Lebedev Physics Institute

Volume 50

ELECTROLUMINESCENCE

Edited by
Academician D. V. Skobel'tsyn
Director, P. N. Lebedev Physics Institute
Academy of Sciences of the USSR, Moscow

Translated from Russian by Albin Tybulewicz
Editor, Soviet Physics - Semiconductors

CONSULTANTS BUREAU
NEW YORK–LONDON
1972

The Russian text was published by Nauka Press in Moscow in 1969 for the
Academy of Sciences of the USSR as Volume 50 of the Proceedings (Trudy)
of the P. N. Lebedev Physics Institute. The present translation is published under an agreement with Mezhdunarodnaya Kniga, the Soviet book ex-
port agency.

Library of Congress Catalog Card Number 71-157932

ISBN 978-1-4757-0315-3   ISBN 978-1-4757-0313-9 (eBook)
DOI 10.1007/978-1-4757-0313-9

© 1972 Consultants Bureau, New York
Softcover reprint of the hardcover 1st edition 1972
A Division of Plenum Publishing Corporation
227 West 17th Street, New York, N. Y. 10011

United Kingdom edition published by Consultants Bureau, London
A Division of Plenum Publishing Company, Ltd.
Davis House (4th Floor), 8 Scrubs Lane, Harlesden, NW10 6SE, London, England

CONTENTS

KINETICS OF THE DESTRIAU EFFECT

E. Yu. L'vova

A description is given of a general method for investigating the kinetics of the Destriau effect. This method is based on the transport equations derived for a crystal phosphor with three systems of levels in the forbidden band: levels of luminescence centers, shallow electron traps, and deep donors. These equations make allowance for all the transitions between each of the systems of levels and both allowed bands, for band — band transitions, as well as for carrier drift in an electric field. The system of transport equations is applied to a two-step model of the Destriau effect. The system of transport equations is simplified considerably by considering the high-field region separately from the recombination region. The proposed theory is used to derive an expression for the electroluminescence efficiency. The results are given of an experimental study of the dependence of the electroluminescence efficiency on the voltage applied to capacitors containing ZnS: Cu phosphors with different concentrations of copper. These experimental results are compared with the theory.

The Destriau effect is the electroluminescence which appears in powdered phosphors when they are excited by alternating fields without making physical contact with a sample [1]. This effect has been studied by many workers. The modern point of view on this effect is when presented consistently in recent monographs (such as that of Henisch [2]) and reviews. Much of the work on the Destriau effect in zinc sulfide phosphors has been done by Soviet physicists. In particular, two reviews on the subject (A. N. Georgobiani in 1963 [3] and Yu. P. Chukova in 1966 [4]) were published in the present serial. A quantitative description of the kinetics of the Destriau effect can now be given from a single standpoint. The purpose of the present paper is to describe a general method for investigating the Destriau effect by means of a system of transport equations and to consider the results obtained by the present author, particularly from experiments involving energy absorption.

CHAPTER 1.

THEORY OF THE KINETICS OF THE DESTRIAU EFFECT

§ 1. General Theory

A. Statement of the Problem. Investigations of electroluminescence usually start with a qualitative description of the process, followed by the derivation of some quantitative characteristic, and a comparison of this characteristic with the results of measurements. Usually, no consideration is given to other dependences or characteristics which follow from the assumptions made to consider a particular property, and no attempt is made to check these other characteristics experimentally.

One of the most important problems in the excitation of a crystal is the mechanism of the formation of a high-field region and the nature of its distribution across a crystal. It is usual to postulate *a priori* that this distribution can be represented by a Mott − Schottky barrier, in which the maximum field intensity is $E \propto \sqrt{V}$, where V is the applied voltage, and the field within the barrier is assumed to decrease linearly with the coordinate. Clearly, this *a priori* representation of the Mott − Schottky barrier and its validity are far from universal.

We shall consider the kinetics of the Destriau effect from a single standpoint. We shall describe the process considered by a system of transport equations for a crystal phosphor with three systems of levels in the forbidden band: levels of luminescence centers, shallow electron traps, and deep donors (Fig. 1). This is the smallest number of systems of levels which can be used in an explanation of the basic experimental observations. The transport equations must make allowance for all possible transitions between each of the levels and both bands, for band − band transitions, and for the drift of electrons and holes in an electric field. Carrier diffusion can be ignored because carriers are transported mainly by drift if the field is $E \sim 10^3$ V/cm, the grains are a few tens of microns across, and the electron mobility is $\mu^- \approx 100$ cm$^2 \cdot$ V$^{-1} \cdot$ sec^{-1} (under these conditions, the electron drift time is 10^{-8} sec).

The mathematical formulation of the problem is based on the kinetics of luminescence considered in Fok's book [5]. This approach is fairly general and Fok's system of equations can be used to describe the electroluminescence of crystal phosphors by introducing terms representing carrier drift in an electric field. In the special case of static fields, the system of equations describes also the dc electroluminescence. Attempts to describe electroluminescence in this way have been made in [6, 7], and the problem of the relationship between the brightnesses of blue and green bands has been tackled in [8]. However, special and incomplete models have been used in all three cases.

A mathematical description of the kinetics of electroluminescence demands a clear and consistent formulation of the basic general ideas on this process. The approach suggested here is free of *a priori* assumptions about the excitation mechanism and, in particular, it is not based on the Mott−Schottky barrier approximation. In principle, this approach can be used to find the nature of the field distribution in a given crystal under conditions.

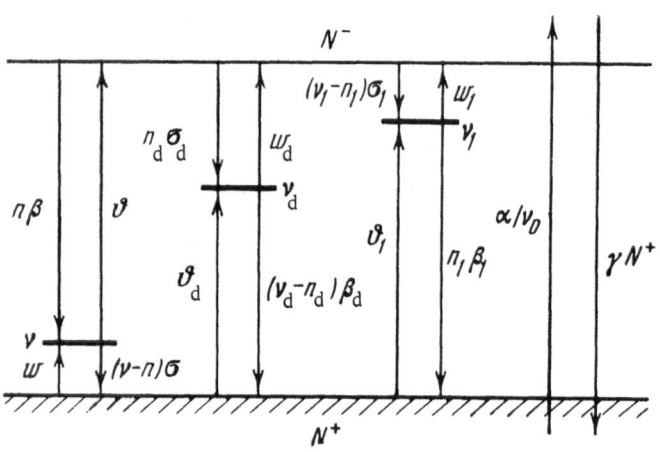

Fig. 1. Energy band scheme of a phosphor crystal with three levels in the forbidden band. The arrows represent electron transitions, and probabilities of these transitions are indicated by letters alongside the arrows; ν, ν_1, ν_0, and ν_d are the total concentrations of, respectively, the luminescence centers, electron traps, lattice atoms, and donors; n, n_1, and n_d are, respectively, the concentration of the ionized luminescence centers, the density of the localized electrons, and the concentrations of the ionized donors; N^- and N^+ are the densities of free charges in the conduction and valence bands.

B. General Transport Equations.
We shall formulate first a basic physical description of a system of levels, which must be used in dealing with electroluminescence, and of the charge states of centers under thermodynamic equilibrium conditions and after the absorption of energy from an external field (excitation).

We shall assume that the luminescence center is neutral before excitation and that it is ionized by excitation, surrendering an electron to the conduction band and becoming positively charged. This is equivalent to the assumption that the luminescence center is a deep donor. It may happen that the nature of the luminescence centers is best represented by a donor – acceptor pair. However, such centers are outside the scope of the present article.

We shall also assume that the electron traps are neutral before their excitation, and that they become negatively charged in the excited state by capturing an electron each from the conduction band. In other words, we shall postulate that such shallow traps are in effect deep acceptors. Thus, the concentration of occupied traps will be simply equal to the density of the localized electrons.

Many experimental observations can be described only if we assume the existence of a third system of levels which are also deep donors but relatively shallower than the luminescence centers. Before their excitation, these donors are assumed to be neutral, and after having been excited, they give up an electron each to the conduction band and become positively charged.

Thus, before their excitation all the centers are neutral (charge 0); after their excitation the luminescence centers and the deep donors become positively charged (charge +1), whereas the shallow traps become negatively charged (charge –1).

The general form of the system of transport equations is

$$\frac{\partial n}{\partial t} = (\nu - n)\,\vartheta - N^- n\beta + N^+ (\nu - n)\,\sigma - nw,$$

$$\frac{\partial n_1}{\partial t} = -n_1 w_1 + N^- (\nu_1 - n_1)\,\sigma_1 - N^+ n_1\beta_1 + (\nu_1 - n_1)\,\vartheta_1,$$

$$\frac{\partial n_d}{\partial t} = (\nu_d - n_d)\,w_d - N^- n_d\sigma_d - n_d\vartheta_d + N^+ (\nu_d - n_d)\,\beta_d,$$

(1)

$$\frac{\partial N^-}{\partial t} = n_1 w_1 + (\nu_d - n_d)\,w_d + (\nu - n)\,\vartheta - N^+ (\nu_1 - n_1)\,\sigma_1 - N^- n_d\sigma_d - N^- n\beta + \alpha - \gamma N^+ N^- - \frac{\partial}{\partial x}(N^- \mu^- E),$$

$$\frac{\partial N^+}{\partial t} = -N^+ (\nu - n)\,\sigma + nw - N^+ (\nu_d - n_d)\,\beta_d + n_d\vartheta_d - N^+ n_1\beta_1 + (\nu_1 - n_1)\,\vartheta_1 + \alpha - \gamma N^+ N^- + \frac{\partial}{\partial x}(N^+ \mu^+ E),$$

$$\frac{\partial E}{\partial x} = \frac{4\pi q}{\varepsilon}(N^+ + n + n_d - n_1 - N^-).$$

The symbols used in these equations are explained in Fig. 1. The quantities μ^- and μ^+ are, respectively, the electron and hole mobilities; q is the electronic charge.

Since the changes in the number of free carriers occur not only because of the trapping at and the liberation from various levels at a given point but also because of the loss (or gain) of carriers resulting from the drift in an electric field, the equations for N^+ and N^- include the drift terms.

The quantities N^-, N^+, n, n_1, and n_d are functions of the coordinates and time and of the transition probabilities. These probabilities are not constant but depend on the field and the field is governed by the space charge of all the centers. Thus, a characteristic feature of electroluminescence is its self-regulation, which is expressed by the self-consistence of the equations in terms of the electric field E.

   The boundary conditions can be specified only for a definite excitation mechanism. If there is no external electron source and, therefore, no electron flow across the boundary of a crystal under ionization conditions, we have the boundary condition

$$N^-_{x=0}\mu^- E_{x=0} = 0. \tag{2}$$

Since the field and electron mobility are not equal to zero, the condition given by Eq. (2) reduces to

$$N^-_{x=0} = 0. \tag{3}$$

   If electrons can reach a crystal from outside, there is a flow of electrons across the boundary of the crystal and the boundary condition can be written thus:

$$N^-_{x=0}\mu^- E_{x=0} = f_0, \tag{4}$$

where the nature of the function f_0 is determined by the mechanism of the electron flow. Thus, for example, if electrons reach the conduction band of ZnS by tunnel transitions across a potential barrier, we find that

$$f_0 \propto \exp\left(-\frac{b_1}{E}\right), \tag{5}$$

if the external voltage is negative at the 'end of a crystal corresponding to x = 0, and $f_0 = 0$ if the external voltage is positive at this point. The coefficient of proportionality between f_0 and the exponential function depends on the external source of electrons. This source may be a second phase, for example, copper sulfide on the surfaces of ZnS grains. The results of an experimental investigation of the role of such a second phase /9/ are presented in Chap. II of the present paper.

   If we assume superbarrier penetration of electrons from outside, we find that the function f_0 becomes

$$f_0 \propto \exp\left(-\frac{q\varphi}{kT}\right), \tag{6}$$

where φ is the height of the potential barrier.

   We shall assume that holes escape from a high-field region to the surface of a crystal. This assumption is correct if free holes do not accumulate in the interior of a crystal. The current of holes which can reach the surface of a crystal of length d can be represented by

$$\int_0^d [nw + n_d\vartheta_d + (v_1 - n_1)\vartheta_1 + \alpha - N^+(v - n)\mathfrak{z} - N^+(v_d - n_d)\beta_d - N^+ n_1\beta_1 - \gamma N^+ N^-] \, dx = N^-_{x=0}\mu^+ E_{x=0}. \tag{7}$$

The condition in Eq. (7) is equivalent to

$$\frac{\partial}{\partial t}\int_0^d N^+(x, t)\, dx = N^+_{x=d}\,\mu^+ E_{x=d}. \tag{8}$$

Equation (8) means that the change in the total number of holes in a crystal is solely due to their penetration from that end of the crystal to which a positive external voltage is applied at any given moment.

   The general system of transport equations gives a mathematical description of the processes which occur in a crystal phosphor grain. Naturally, the complete solution of such a system is exceptionally difficult. Therefore, the transport equations must be studied under specific conditions.

C. Transport Equations for High-Field and Recombination Regions.
We shall now consider some specific and yet fairly general representations of the Destriau
effect which can be used to simplify the problem.

First, we shall assume that electroluminescence excited by an alternating voltage is a
two-step process. This assumption is supported by many experiments and may be regarded as
fully established. In other words, we shall assume that the excitation of a crystal and the
conversion of such excitation into radiation can be regarded as being separated in space
(occurring in different parts of a grain) and time (occurring at different periods of the exciting
alternating voltage). Therefore, we shall consider separately that region of a grain in which a
strong field is established and luminescence centers are ionized during a given half-period of
the voltage. We shall then consider a different region in a grain where carrier recombination
takes place.

Secondly, we shall assume that during electroluminescence the field is concentrated in a
certain region, as demonstrated in many experiments. The nature of the space charge responsible for the establishment of a high-field region will be considered on the basis of general representations of the energy band structure. No *a priori* assumptions will be made about the field
distribution.

Finally, we shall take into account the law of conservation of charge and make allowance
for the drift of carriers in an electric field.

Before we derive a system of equations for the high-field region, we must consider the
nature of the space charge which is responsible for the concentration of the field in the ionization (high-field) region.* If the energy of the external field is sufficient to ionize a luminescence
center separated by 3.0 eV from the edge of the conduction band, this energy would be quite
sufficient to transfer an electron from the valence band to the level of a luminescence center
(about 0.8 eV is required), and this would result in the deionization of the luminescence
center.†

In fact, the probability of the liberation of an electron from a local level by an external
field depends exponentially on the field intensity and the depth of the levels. Therefore, a
severalfold change in the depth of the level results in a change of the liberation probability
amounting to several orders of magnitude. It follows that the probability of liberation of an
electron ϑ from a luminescence center is much less than the probability of an electron
transition from the valence band to the luminescence center: $\vartheta \ll \omega$. This means that, under
steady-state conditions, we have $n \ll \nu - n$. Thus, the charge does not become localized at the
luminescence centers in the high-field region.

Exactly the same considerations apply to shallow electron traps, i.e., there is no
accumulation of charge at these traps. The levels of these deep acceptors may be located both
above and below the donor levels.

In our model, the field in the excitation (ionization) region is due to a positive space
charge. The levels at which this charge is stored are located in the upper half of the forbidden
band and, therefore $w_d > \vartheta_d$. It follows that, under steady-state conditions, we have $n_d \gg$
$\nu_d - n_d$. Consequently, a positive space charge is stored at the donor levels.

---

* We shall consider the specific case of ZnS but the approach can be used also to describe the
  Destriau effect in other phosphors.
† These are the "optical" gaps between the levels because the transitions considered are rela-
  tively fast and satisfy the Franck − Condon principle.

It is natural to assume that there is no strong field in the recombination region ("strong" means here a field sufficient for ionization); however, a weak field capable of causing carrier drift may exist in this region. If the charge of electrons arriving in the recombination region from the opposite end of a crystal is not compensated, it will gradually accumulate and eventually give rise to a strong field. This difficulty can be avoided by assuming that holes liberated by deionization from the luminescence centers are carried away by the field to the other end of the crystal and are trapped at the surface. This assumption results in an important conclusion that holes are generated in the ionization region. During the next half-period, when the sign of the external field is reversed, the holes return from the surface of a crystal (or a grain) to the levels of the luminescence centers and ionize these centers. Electrons arrive from the opposite end of the crystal (or the grain) and recombine with these ionized centers. The role of the holes in the kinetics of electroluminescence has been considered in [10].

We shall not consider the initial stages, i.e., we shall exclude from our consideration the moments corresponding to zero voltage. Our discussion will begin at a certain moment when the field is sufficiently high but has not reached its maximum value. We thus find that the ionization of the luminescence centers and other levels occurs at one end of a crystal (or a grain), holes escape from the luminescence centers to the surface, and electrons are carried away by the field to the other end. The space charge is concentrated at the deep donor levels.

A simple estimate confirms our assumption that free charges are removed almost instantaneously from the region where they are generated. It is shown in [1] that the field in the ionization region may reach 10^6 V/cm. If the electron mobility is of the order of 100 cm^2 · V^{-1} · sec^{-1} and the dimensions of the high-field (ionization) region are about 10^{-4} mm, the residence time of an electron in this region is only $\tau \approx 10^{-12}$ sec. This very low value of τ is obtained also for holes even if their mobility is less than that of electrons.

We shall now write our assumptions in the form of inequalities so as to include them in the simplification of the general system of transport equations.

1. The space charge is concentrated at the donor levels and no charge is stored at the luminescence centers or at the shallow traps. This means that $n \ll n_d$ and $n_1 \ll n_d$. It follows that $n \ll \nu$ and $n_1 \ll \nu_1$.

2. Free electrons and holes are removed instantaneously by the field to different ends of a crystal (or a grain) and form a surface charge of the appropriate sign. This means that $N^- \ll n_1$ and $N^+ << n$. It follows that $N^-, N^+ \ll n_d$ and $N^-, N^+ \ll \nu, \nu_1$.

Thus, the following inequalities are satisfied in the high-field region:

$$N^- n \beta \ll (\nu - n)\, \vartheta,$$
$$N^- n_d \mathfrak{s}_d \ll (\nu_d - n_d)\, w_d , \qquad (9)$$
$$N^- (\nu_1 - n_1)\, \sigma_1 \ll w_1 n_1,$$

as well as the inequalities

$$N^+ (\nu - n)\, \mathfrak{s} \ll nw,$$
$$N^+ (\nu_d - n_d)\, \beta_d \ll n_d \vartheta_d , \qquad (10)$$
$$N^+ n_1 \beta_1 \ll (\nu_1 - n_1)\, \vartheta_1.$$

Moreover,

$$\gamma N^+ N^- \ll \alpha. \qquad (11)$$

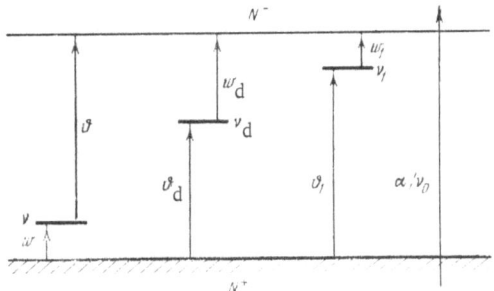

Fig. 2. Energy band scheme of the high-field region in a crystal phosphor.

All these assumptions can be represented graphically by leaving only those transitions which are denoted by arrows pointing in the upward direction (Fig. 2).

If we make the assumptions expressed mathematically by all these inequalities, we find that the general system of equations (1) reduces to

$$\frac{\partial n}{\partial t} = v\vartheta - nw,$$

$$\frac{\partial n_1}{\partial t} = - n_1 w_1 + v_1 \vartheta_1,$$

$$\frac{\partial n_d}{\partial t} = (v_d - n_d) w_d - n_d \vartheta_d,$$

$$\frac{\partial N^-}{\partial t} = n_1 w_1 + (v_d - n_d) w_d + v\vartheta + \alpha - \frac{\partial}{\partial x}(N^- \mu^- E),$$ (12)

$$\frac{\partial N^+}{\partial t} = nw + n_d \vartheta_d + v_1 \vartheta_1 + \alpha + \frac{\partial}{\partial x}(N^+ \mu^+ E),$$

$$\frac{\partial E}{\partial x} = \frac{4\pi q}{\varepsilon} n_d.$$

Such a simplification of the band transition scheme and of the corresponding system of equations for the high-field region is strictly valid only if our three general physical assumptions are justified.

The boundary conditions for electrons remain the same as for the general system of equations [see Eqs. (2)-(6)]. The boundary condition for holes simplifies to

$$\int_0^{L(t)} (nw + n_d \vartheta_d + v_1 \vartheta_1 + \alpha) \, dx = N^+_{x=0} \mu^+ E_{x=0},$$ (13)

where L(t) is the boundary of the high-field region. This condition is satisfied if the holes from the recombination region do not penetrate the high-field region.

The system of equations for the high-field region can be transformed so that its solution can be expressed in terms of the quadrature formulas. Thus, if we know explicitly the dependences of the various quantities on the field E, the solution can be obtained by means of relatively simple numerical calculations.

The same assumptions can be applied to the recombination region. In this case, we leave only those transitions which are denoted by the arrows pointing downward (Fig. 3), provided the levels of the luminescence centers and electron traps are sufficiently deep. The system of transport equations now becomes

$$\frac{\partial n}{\partial t} = - N^- n\beta + N^+ (v - n) \sigma,$$

$$\frac{\partial n_1}{\partial t} = N^- (v_1 - n_1) \sigma_1 - N^+ n_1 \beta_1,$$

$$\frac{\partial n_d}{\partial t} = - N^- n_d \sigma_d + N^+ (v_d - n_d) \beta_d.$$

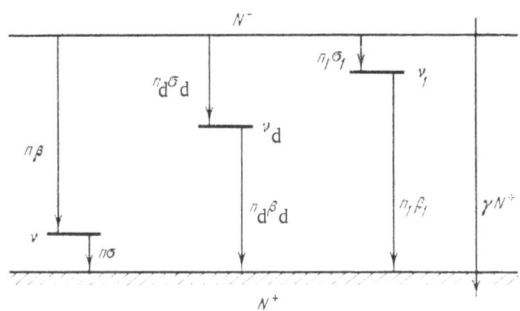

Fig. 3. Energy band scheme of the recombination region in a crystal phosphor.

$$\frac{\partial N^-}{\partial t} = - N^-(\nu_1 - n_1)\,\sigma_1 - N^- n_d \sigma_d - N^+ n_1 \beta_1 + \frac{\partial}{\partial x}(N^- \mu^+ E),$$

$$\frac{\partial N^+}{\partial t} = - N^+(\nu - n)\,\sigma - N^+(\nu_d - n_d)\,\beta_d - N^- n\beta - \frac{\partial}{\partial x}(N^- \mu^- E),$$

$$\frac{\partial E}{\partial x} = \frac{4\pi q}{\varepsilon}(n + n_d + n_1). \tag{14}$$

In writing down the boundary conditions, we shall make allowance for the fact that the electron current reaching the boundary of the recombination region L (t) consists of electrons liberated from all the levels in the ionization region:

$$\int_0^{L(t)} [(\nu - n)\,\vartheta + (\nu_d - n_d)\,w_d + n_1 w_1 + \alpha]\,dx = N^-_{x=L(t)}\,\mu^- E_{x=L(t)}. \tag{15}$$

Since the charge does not accumulate in the recombination region, the electron and hole currents reaching this region should be equal:

$$N^+_{x=d}\mu^+ E_{x=d} \approx N^-_{x=L(t)}\,\mu^- E_{x=L(t)}. \tag{16}$$

This discussion of two regions in a crystal in which different physical phenomena are taking place, i.e., the use of the two-step model of electroluminescence, requires the introduction and consideration of some boundary L(t) between these regions. This boundary defines the high-field region. Thus, our treatment describes quite satisfactorily the phenomena occurring outside the region L(t) but we cannot say anything definite about the processes occurring in the immediate vicinity of the boundary of this region. This boundary is defined by the relationship between the external potential difference V(t) and the field in a crystal (or a grain):

$$\int_0^{L(t)} E(x, t)\,dx = - V(t). \tag{17}$$

The voltage drop in the recombination region can be neglected.

We shall conclude this section by pointing out that our general approach to the kinetics of electroluminescence makes it possible to refine some of our ideas on its mechanism without making explicit assumptions. Thus, for example, it follows from our analysis that the positive space charge, which determines the concentration of the field, should be located at fairly deep donor levels. Poisson's equation in the form given by Eq. (12) is obtained only when it is assumed that the ionization and recombination regions can be considered separately and the stated assumptions are made about the nature of the space charge and about the motion of free charges in the field. However, this is insufficient to obtain the field distribution typical of a Mott — Schottky barrier: it is necessary to assume also that the concentration of ionized donors in the barrier region is constant. If this assumption is not made, Poisson's equation (12) can be used to find the field distribution which is frequently closer to the true situation than that given by Mott − Schottky barrier model.

The formation of ionized luminescence centers is usually considered to be simply the result of excitation and, in many cases, the details of this process are unimportant. In a unified description of the kinetics, we have to assume that holes appear first in the excitation region and that they are then removed by the field to the surface of the crystal (or the grain) and that only these holes return to the levels of the luminescence centers during the next half-period of the exciting voltage. These hole processes are usually ignored but they may be important, for example, if the nature of the polarization field in a crystal is considered.

By way of example, we shall consider the assumptions used in investigations of the phase of the principal peak of a brightness wave. It is assumed in [12] that the brightness is proportional to the number of electrons lost from the trapping centers and, therefore, that the time dependence of the brightness is described by the equation (reformulated using our symbols)

$$B(t) \propto \frac{\partial n_1}{\partial t}, \qquad \frac{\partial n_1}{\partial t} = -w_1 n_1. \tag{18}$$

This implies that all our three principal assumptions (employed in the simplification of the equations for the high-field region) are used. Moreover, it is sufficient to consider only the electron transitions to the conduction band; the hole transitions from the trapping levels to the valence band can be ignored (Fig. 2).

We shall mention also an additional feature of electroluminescence. The energy absorbed in electroluminescence is acquired directly from an electric field. The energy of an external field can be absorbed only if charges move in this field. In contrast to other types of luminescence (photoluminescence, cathodoluminescence, x-ray luminescence, etc.), for which the energy absorbed by a phosphor is governed by the power of the source and is independent of the properties of the phosphor, the amount of energy acquired from the field is determined not so much by the power of the excitation source as by the state of the crystal. If, for example, a crystal does not contain free charges or charges which can be liberated easily, it cannot simply absorb the energy from an external field. The number of free charges depends on the state of excitation of the phosphor and this, in turn, is governed by the absorbed energy. In this sense, electroluminescence is a self-regulating process and the controlling property is the number of electrons participating in the electroluminescent processes. This property can be seen very clearly from the transport equations.

§2. Derivation of Formulas for Electroluminescence
Efficiency from the Transport Equations

Electroluminescence can be investigated in two ways: by studying the luminescence characteristics, or the electrical properties. The first approach has been used by many authors and studies of the properties of luminescence (brightness waves and the average brightness during one period of the applied voltage) have served as a basis for the ideas on the mechanism of electroluminescence developed by Georgobiani and Fok [11-18].

Investigations of the absorption of the energy of external excitation is the second approach to studies of the kinetics of electroluminescence. This way is of intrinsic interest in elucidating the factors which cause the energy of external excitation to be lost in a phosphor crystal. A characteristic feature of luminescence processes is the existence of intermediate stages within a luminescent body between the absorption of the energy of external excitation and the emission of a photon. Thus, studies of the absorption of energy, of the processes is which this energy is utilized, and of the causes of the energy losses are essentially equivalent to studies of the conversion of energy within a phosphor.

Energy may be lost at any stage of its conversion from the energy of an external field to the energy of the emitted photon. Although the motion of charges in an electric field is always accompanied by the absorption of energy, it does not always result in the ionization or excitation of luminescence centers. It may happen also that ionized luminescence centers may become de-ionized by a nonradiative process. The energy losses may thus occur before and after the ionization of luminescence centers.

We shall study the kinetics of the Destriau effect and check current ideas on the processes governing the brightness by using the second approach, i.e., by measuring the energy absorbed by a phosphor as a function of the applied voltage. Such a study of energy absorption is also important in the determination of the efficiency of electroluminescent light sources and other devices.

Our theoretical treatment can be used to derive an expression for the efficiency of electroluminescence.

We shall find first the expression for the dependence of the period-average electroluminescence brightness $\overline{B}$ on the applied voltage. We shall assume that this brightness is equal to the total number of recombination events. Under steady-state conditions the number of ionization events is equal to the number of recombination events and therefore

$$\overline{B} = \frac{1}{T} \int_0^T \int_0^{L(t)} (v\vartheta + \alpha)\, dx\, dt = \frac{1}{T} \int_0^T \int_{L(t)}^d N^- n\beta\, dx\, dt, \qquad (19)$$

where T is the period of the applied external voltage.

Following our initial assumption that electroluminescence is a two-step process, we shall postulate that ionization occurs only in the high-field region where the ionization probability is exceptionally high.

Furthermore, we shall assume that the process of establishment of the excited state in a phosphor crystal takes place in two stages: electrons from an external source reach the high-field region where they acquire sufficient energy and impact-ionize the crystal lattice and the luminescence centers.

The number of ionization events involving luminescence centers is proportional to the number of free electrons N^- which penetrate the high-field region from an external source, to the number of un-ionized luminescence centers, and to their impact ionization probability. We have demonstrated in connection with the transport equations that $n \ll \nu - n$. The impact ionization probability depends on the field in accordance with the law [19, 20]

$$w_i \propto \exp\left(-\frac{b}{E^2}\right), \qquad (20)$$

and, therefore,

$$v\vartheta \propto v N^- \exp\left(-\frac{b}{E^2}\right). \qquad (21)$$

If the probability of ionization of the crystal lattice depends on the field in a similar manner, it follows that the number of lattice ionization events is

$$\alpha \propto v_0 N^- \exp\left(-\frac{b'}{E^2}\right), \qquad (22)$$

where ν_0 is the number of lattice atoms ionized per unit volume.

The transport equations for the high-field region [the fourth equation in the system (12)] can be used to find the value of N^-. Since electrons are removed practically instantaneously from the high-field region, it follows that the absolute value of their density must be low. Hence, the absolute rate of change of the electron density is low compared with the rate of change of the total number of electron liberation events:

$$\frac{\partial N^-}{\partial t} \ll (v_d - n_d)\, w_d \quad v\vartheta, \ \alpha. \tag{23}$$

We shall also assume that the shallow levels are completely ionized at the moment of formation of the high-field region. Therefore, the term $n_1 w_1$ in the fourth equation of the system (12) can be ignored and this equation becomes

$$\frac{\partial}{\partial x}\left(N^-\mu^- E\right) = (v_d - n_d)\, w_d + v\vartheta + \alpha. \tag{24}$$

We shall use the function $f = N^-\mu^- E$. This function represents the number of electrons crossing a unit area per unit time. It depends on x and t.

The boundary condition for the last equation

$$\frac{\partial f}{\partial x} = (v_d - n_d)\, w_d + v\vartheta + \alpha \tag{25}$$

follows from the assumption that the source of electrons is external (for example, the second phase in a crystal). Electrons penetrate a crystal by tunneling, and the probability of a tunnel transition depends on the field in accordance with the law [21]:

$$w_t \propto \exp\left(-\frac{b_1}{E}\right), \tag{26}$$

the boundary condition being

$$f_0(t) = N^-_{x=0}\,\mu^- E_{x=0} = A_0 \exp\left(-\frac{b_{1,0}}{E_{x=0}}\right). \tag{27}$$

We shall solve our equation bearing in mind that the ionization probability depends strongly on the field and, therefore, on the coordinate. Consequently, we may assume that most of the energy needed for the ionization is acquired by the electrons in a narrow part of the high-field region near the boundary of the crystal at $x = 0$. This means that

$$f(x, t) \approx f_0(t). \tag{28}$$

Then

$$N^- = \frac{f_0}{\mu^- E_{x=0}}. \tag{29}$$

Calculating the integral on the left-hand side of Eq. (19) by the method of steepest descents, we obtain

$$\int_0^{L(t)} (v\vartheta + \alpha)\, dx = v\vartheta + \alpha\big|_{x=0}\, \Delta x = \frac{f_0 \exp\left(-\dfrac{b_{1,0}}{E_{x=0}}\right)\Delta x k}{\mu^- E_{x=0}} \times$$

$$\times \left[v \exp\left(-\frac{b}{E_{x=0}}\right) + k' v_0 \exp\left(-\frac{b'}{E^2_{x=0}}\right)\right] \quad (k, k' = \text{const}), \tag{30}$$

where Δx is the distance from that boundary of the crystal at which the electrons acquire the energy needed in the act of ionization. Formulas (21) and (22) hold for distances $\Delta x \ll L(t)$. We shall next calculate the integral with respect to t by the method of steepest descents and assume $E \propto \sqrt{V}$ in a Mott – Schottky barrier. This gives the following expression for the dependence of the average brightness on the voltage:

$$\overline{B} = \frac{A_0 \exp\left(-\dfrac{b_1}{\sqrt{V}}\right)\Delta x \Delta T k}{T\mu^- \sqrt{V}}\left[\nu\exp\left(-\frac{b}{V}\right)+k'\nu_0\exp\left(-\frac{b'}{V}\right)\right]. \tag{31}$$

An analysis of the transport equations thus yields an expression for the electroluminescence brightness, which can be compared directly with the experimental data.

It follows from Eq. (31) that the brightness is an exponential function of the voltage and that it is proportional to the product of two exponential functions with different powers of the voltage in their arguments:

$$\overline{B} \propto \exp\left(-\frac{b_1}{\sqrt{V}}\right)\exp\left(\frac{b}{V}\right), \tag{32}$$

(the quantities Δx and ΔT are independent of the voltage).

Experience shows that, over a wide range of voltages and for values of the brightness ranging over many orders of magnitude, the dependence $\overline{B}(V)$ obeys quite accurately the law

$$\overline{B} \propto \exp\left(-\frac{b_1}{\sqrt{V}}\right) \tag{33}$$

(Fig. 4). Therefore, we may assume that $\exp\left(-\dfrac{b}{V}\right) \approx 1$. (Usually, the expression

$$\left[\nu\exp\left(-\frac{b}{V}\right)+\nu_0\exp\left(-\frac{b'}{V}\right)\right]\Big/\sqrt{V}$$

is approximated by a power function V^n with $n = 1-2$.) It must be pointed out, however, that although the dropping of the factor $\exp\left(-\dfrac{b}{V}\right)$ is justified for ZnS, it may be incorrect for other phosphors. In fact, this factor may play a dominant role, as demonstrated by Vereshchagin and Drapak [22].

The value of the electroluminescence efficiency is governed by the relative role of two factors which determine the brightness: the number of electrons which participate in electroluminescence, and the conditions which exist during their motion (acceleration) in the internal electric field whose spatial and time distributions may be quite complex. In theory, we should consider both factors but in practice one of them predominates. If the acceleration conditions play the dominant role, the electric field energy is absorbed by many electrons but this energy is transferred to the luminescence centers only by those electrons which reach a velocity sufficiently high for the ionization of these centers. The other

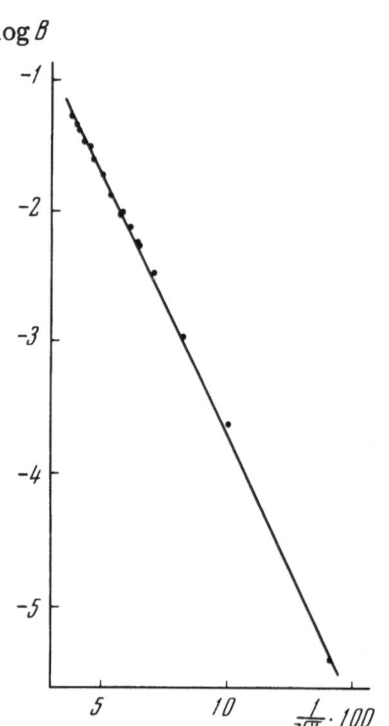

Fig. 4. Dependence of the electroluminescence brightness on the voltage.

electrons transfer their energy to the thermal vibrations of the crystal lattice. In this case, even the theoretical value of the electroluminescence efficiency cannot be high. On the other hand, if the dominant factor is the number of electrons and only a small number of electrons reaches the high-field region but almost all of them acquire energy sufficient for the ionization of the lattice or the luminescence centers, the efficiency may be high. It follows that a search for the conditions most favorable for the dominance of this factor is justified. The experiments of Georgobiani and Fok demonstrate that the dominant factor in the electroluminescence of ZnS is the number of electrons and not the acceleration in an electric field. Therefore, the nature of the dependence of the average brightness on the voltage does not change at the critical value of the field [15]. The experiments of Bukke, Vinokurov, and Fok [23] also confirm that the decisive factor is the number of electrons taking part in the electroluminescence processes. They showed that the introduction of a single additional electron results in the emission of from 20 to 100 photons.

We shall now use our equations to derive a formula for the absorbed energy.

The instantaneous power absorbed by a phosphor crystal is

$$W(t) = S \int_0^d \Phi E dx, \tag{34}$$

where Φ is the total flux of positive and negative charges; S is the cross section of the crystal. The energy of the external field may be absorbed by electrons and holes. Therefore, in general, we find that

$$\Phi = - q N^- v^- + q N^+ v^+ \tag{35}$$

(v^- and v^+ are the electron and hole velocities, respectively). The general expressions for N^- and N^+ can be found from the system of transport equations (1).

Equation (34) can be simplified and reduced to a form which can be checked experimentally provided the flow of charge through a unit area in a grain is constant. The constancy of this flow of charge may be realized if (1) the number of positive and negative charges liberated per unit time is equal at each point in the high-field region after the formation of this region, and (2) the energy is absorbed mainly by electrons liberated at the boundary of the high-field region so that they are subjected to practically the whole of the potential applied across a grain. A Mott-Schottky barrier should then form in the high-field region provided the impurities are distributed uniformly throughout the grain. In this case, we obtain

$$W(t) = - S \int_0^d q N^- v^- E dx. \tag{36}$$

Using the function f(x, t) = $N^- \mu^- E$, introduced earlier, and the relationship $v^- = E \mu^-$, we obtain

$$W(t) = - q S \int_0^d f(x, t) E dx. \tag{37}$$

Most of the voltage drop across a grain is concentrated in the high-field region. Therefore, we may assume that the energy of an external source is absorbed mainly in the high-field region and that the ohmic losses elsewhere in the grain are relatively small. We thus find that the voltage applied across a grain is almost completely concentrated in the barrier region. Consequently, we need not carry out integration over the whole length of the grain but only up to the boundary of the high-field region:

$$W(t) = - q S \int_0^{L(t)} f(x, t) E dx. \tag{38}$$

We shall now replace the function f(x, t) by its average value over the high-field region, $\bar{f}$ (t).

Since $\int_0^{L(t)} E(x, t)\,dx = -V(t)$, Eq. (38) now becomes

$$W(t) = qS\,\bar{f}(t)\ V(t). \tag{39}$$

The strength of the source of electrons is given by the function $\bar{f}$ (t). The absorption of energy is governed by the electrons liberated from the luminescence centers and from the deep traps. At the moment of establishment of a high-field region in a crystal, the shallow traps are already empty and cannot act as the source of energy-absorbing electrons. The greatest role in the energy absorption is played by those electrons which can be liberated after the formation of a high-field region, i.e., when the applied potential difference is fairly high. In other words, we may assume that the energy is absorbed mainly by those electrons which are liberated by the field from fairly deep levels. These levels should be deeper than those from which the electrons are liberated to form a positive space charge. It is shown in [12, 14] that such levels exist in ZnS crystals and are separated by about 0.7 eV from the conduction band. Experiments confirm that when the applied voltage is increased, electrons are liberated from levels of increasing depth right down to the deepest levels which are emptied by a field designated critical [14].

Our analysis of the role of various processes in the absorption of energy by a phosphor leads to the following expression for the function $\bar{f}$ (t), which follows from the system of transport equations (12) for the high-field region,

$$\bar{f}(t) = [v\vartheta + \alpha + (v_d - n_d)\,w_d]\,\Delta x, \tag{40}$$

where Δx is the effective thickness of the region in which the liberation of electrons takes place.

We thus take into consideration the electrons liberated by the ionization of the luminescence centers and of the crystal lattice, as well as the electrons liberated by the field from the deep levels.

The field dependence of the probability of each of these processes can be given explicitly only if we make some assumptions about the mechanism of electron liberation. The number of electrons liberated by the ionization of the lattice and of the luminescence centers, $v\vartheta + \alpha$, has already been found [see Eqs. (30)–(33)]. The liberation of electrons from the local levels occurs (at T > 250°K) by tunnel transitions assisted by the thermal vibrations of the lattice (this was established in a study of the temperature dependence of the critical field). The calculations reported in [24] show that the tunneling probability depends on the field, in accordance with the law

$$w \propto \exp b_2 E^2. \tag{41}$$

We shall now derive an expression for the average power absorbed by a phosphor in one period:

$$\overline{W} = \frac{1}{T}\int_0^T W(t)\,dt. \tag{42}$$

Using Eqs. (39)–(42) and integrating with respect to t by the method of steepest descents, we obtain

$$\overline{W} = \frac{qS}{T}\int_0^T \bar{f}(t)\,V(t)\,dt = \frac{qS\Delta x\Delta T V}{T}\,[v\vartheta + \alpha + (v_d - n_d)\,w_d] = qV\left[A_1\exp\left(-\frac{b_1}{\sqrt{V}}\right) + A_2\exp b_2 V\right] \tag{43}$$

where A_1 and A_2 are constants, which are assumed to depend weakly on V.

Having derived expressions for the brightness (31) and for the power absorbed by a phosphor (43), we can now give the formula for the electroluminescence efficiency as a function of the applied voltage:

$$\rho = \frac{\bar{B}}{\bar{W}} = \frac{A \exp\left(-\frac{b_1}{\sqrt{V}}\right)}{qV\left[A_1 \exp\left(-\frac{b_1}{\sqrt{V}}\right) + A_2 \exp b_2 V\right]}, \quad A = \text{const.} \qquad (44)$$

CHAPTER II

EXPERIMENTAL INVESTIGATION OF THE KINETICS OF THE DESTRIAU EFFECT

§ 1. Investigation of the Dependence of the Electroluminescence Efficiency on the Voltage

Georgobiani, L'vova, and Fok [25-27] investigated ZnS:Cu electroluminescent capacitors used widely in experimental work and in electroluminescent devices. These capacitors were excited by a sinusoidal voltage of 50 Hz frequency (line voltage). The brightness was measured by an FÉU-19 photomultiplier [25] calibrated in absolute energy units for light of definite wavelengths. The power absorbed by an electroluminescent capacitor was measured by the method of direct amplification of the instantaneous values of the current through the capacitor and the voltage across it; this was done using the power vibrator of a loop oscillograph. The investigation method and the capacitor were described in detail in [25].

Measurements of the absolute energy efficiency, carried out on capacitors filled with ZnS:Cu phosphors obtained from different sources demonstrated that the maximum efficiency was about 1.3%. This corresponded to a relative luminous efficiency of about 6 1m/W for green light.

As pointed out earlier, a study of the absorption of energy is one of the two main methods for investigating the mechanism of electroluminescence. Moreover, the experimentally determined dependence of the efficiency on the voltage could be compared with the dependence derived from the general transport equations.

We shall start from the assumption that the energy was absorbed mainly in the phosphor itself. Moreover, the voltage across a given grain will be assumed to be proportional to the external voltage.

The dependence of the efficiency on the voltage, given by Eq. (44), can be transformed to an expression which was more convenient in comparisons with the experimental data:

$$\rho(V) = \frac{a}{V\left[1 + C \exp\left(b_2 V + \frac{b_1}{\sqrt{V}}\right)\right]}, \qquad (45)$$

where $a = A/qA_1$, $C = A_2/A_1$. Since the number of electrons liberated from the donor levels is considerably greater than the number of electrons generated by the ionization of the crystal lattice and the luminescence centers, the first term in the denominator of Eq. (44) or the unity in the denominator of Eq. (45) can be ignored.

In general, the coefficient in front of the exponential function depends weakly on the voltage because the pre-exponential factor in the expression for the probability of the ionization of the donor levels by the field may be a function of the field. This dependence is usually assumed to be linear [24]. We shall postulate that $A_2 \propto \sqrt{V}$ because this gives the best agreement between theory and experimental. Eq. (45) then assumes the form

$$\rho(V) = \frac{a_1}{V^{3/2} C_1 \exp\left(b_2 V + \frac{b_1}{\sqrt{V}}\right)}, \tag{46}$$

where the quantities a_1 and C_1 no longer depend on V.

The influence of scale factors can be eliminated by representing the dependence of the efficiency on the voltage in terms of logarithms

$$\log \rho = -0.43 \left(\frac{3}{2} \ln V + b_2 V + \frac{b_1}{\sqrt{V}}\right). \tag{47}$$

A comparison of theory with experiment thus requires a knowledge of two parameters, b_1 and b_2. Both can be found by independent experiments.

The coefficient b_1 can be found quite accurately from the dependence of the brightness on the voltage $\log B = f(1/\sqrt{V})$ because this dependence is linear in a wide range of values of the voltage and the brightness. Figure 4 shows this dependence for an electroluminescent capacitor filled with a phosphor supplied by the Physics Institute of the Academy of Sciences (we shall denote this phosphor by PIAS).

The coefficient b_2 can also be found from independent measurements of the absorption of energy as a function of the voltage. However, the accuracy of the value of b_2 is low and, therefore, the points on the graph shown in Fig. 5 do not fit a straight line too well. Consequently, the experimental dependence can be represented only approximately by a straight line.

The function $\log W/V^{3/2} = f(V)$ deviated from linearity (Fig. 5) because only one group of trap levels was allowed for in our analysis. In fact, it was known that traps of different depths were present in a phosphor and these were represented by many local levels in the forbidden band. The shallow traps were emptied at lower voltages and, therefore, the dependence $\log W/V^{3/2} = f(V)$ was steeper at low values of V. Moreover, we ignored the absorption of energy by the electrons liberated from the luminescence centers, because we dropped unity from the denominator of Eq. (45). In fact, this term was of some importance and its influence was particularly strong near the efficiency maximum [more precisely, near the minimum of the exponential function $\exp(b_2 V + b_1/\sqrt{V})$], whereas at low and high values of V this term was not important.

To determine the parameter b_2 more accurately, we plotted a family of theoretical curves corresponding to different values of b_2 in Eq. (47) for each of the investigated capacitors (this was done using the experimentally determined value of b_1). These theoretical curves were compared with the experimental dependences of the efficiency on the voltage. The theoretical curve was made to coincide with the experimental data solely by a shift along the ordinate. The results of such a comparison are presented in Fig. 6. It is evident from this figure that the best agreement between theory and experiment near the efficiency maximum and in the range of high voltages was obtained for $b_2 = 0.006V^{-1}$ for the PIAS phosphor and for $b_2 = 0.005 V^{-1}$ for the phosphors obtained from the State Institute of Applied Chemistry (SIAC) and from Czechoslovakia.

These values of the parameter b_2 were compared with the values deduced from the dependence of the absorbed energy on the voltage by plotting the straight lines in Fig. 5 with slopes given by the newly found values of b_2. It is evident from Fig. 5 that the slopes of the calculated lines were in agreement with the experimental dependences.

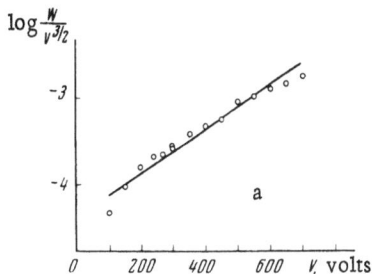

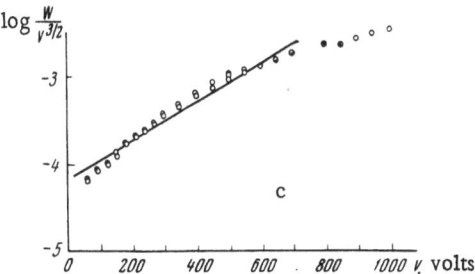

Fig. 5. Dependence of the power on the voltage across electroluminescent capacitors containing ZnS:Cu phosphors: a) PIAS phosphor; b) phosphor from Czechoslovakia; c) SIAC phosphor. The continuous lines are theoretical; they are calculated for the following values of the parameter b_2: a) $0.006 V^{-1}$; b), c) $0.005\ V^{-1}$. The open circles represent the experimental values.

The coefficient b_2 can be estimated also by a third method: it can be found from the temperature dependence of the critical field [14]. It was reported in [14] that the room-temperature value of the critical voltage for our capacitors was $V_{cr} = 600$ V. The depth of the donor levels was assumed to be 0.67 eV. These results showed that the values of b_2 should be $0.01 \pm 0.004\ V^{-1}$. Since the thickness of the phosphor layer could vary somewhat from one capacitor to another (but the variation should not be greater than a factor of 1.5), the range of b_2 was effectively $0.01 \pm 0.006\ V^{-1}$. The values of the parameter b_2 found by the other two methods were thus within the range just quoted.

The mechanism of electroluminescence suggested on the basis of studies of the emitted radiation was thus confirmed by experiments concerned with the voltage dependence of the efficiency, and the formula for the efficiency deduced from transport equations was confirmed experimentally.

§2. Influence of the Activator Concentration and of a Second 'Phase' on the Energy Absorption and the Electroluminescence Brightness

The excited state can be established in a crystal only if it contains free charges capable of absorbing the energy of an external electric field. Therefore, the prime requirement is a source of electrons. Next, the field must be concentrated in the crystal in such a way that these electrons acquire energies sufficient for the ionization of the luminescence centers. It follows that the electroluminescence brightness is governed by two factors: the strength of the source of electrons and the field concentration conditions.

We shall assume that the average electroluminescence brightness is governed by the tunneling of electrons through a potential barrier and that the source of these electrons may be a second phase (for example, copper sulfide) located at the boundaries phosphor grains [15].

The idea that electroluminescence may be associated with the formation of a second phase follows from the data on the preparation of electroluminescent phosphors. It is known

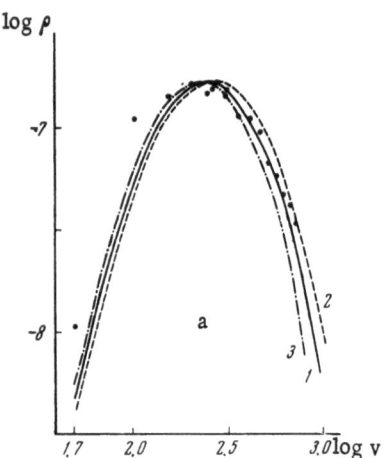

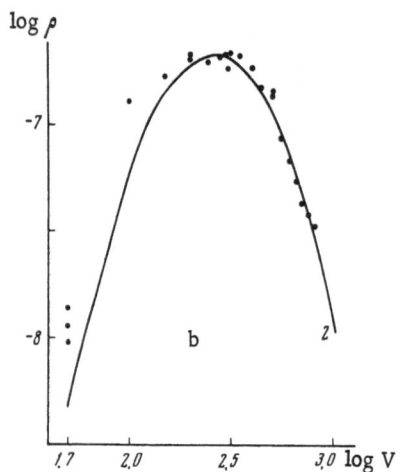

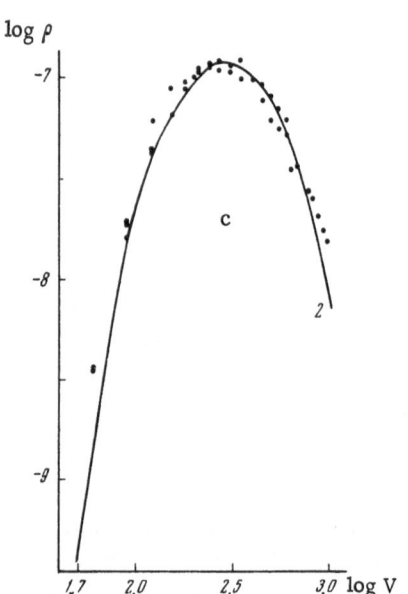

Fig. 6. Dependence of the electroluminescence efficiency on the voltage across electroluminescent capacitors containing ZnS:Cu phosphors: a) PIAS phosphor; b) phosphor from Czechoslovakia; c) SIAC phosphor. 1) $b_2 = 0.006$ V^{-1}; 2) $b_2 = 0.005$ V^{-1}; 3) $b_2 = 0.007$ V^{-1}. The experimental results are represented by points and the theoretical calculations by dashed, and chain curves.

that considerably higher copper concentrations (of the order of $10^{-4} - 10^{-3}$ g-atom of Cu per mole of ZnS) are needed for efficient electroluminescent phosphors than for photoluminescent materials. Electroluminescent phosphors are usually prepared by heating in a mixture of H_2S and HCl gases at about 1000°C. Under these conditions, some fraction of copper penetrates the crystal lattice and forms the luminescent ZnS:Cu phase, whereas the excess copper (which has a limited solubility) forms a second phase of copper sulphide at the grain boundaries. We shall not consider the chemical nature and spatial distribution of the second phase in detail, and shall bear in mind that the concept of a phase is itself somewhat arbitrary. We shall be interested simply in the existence of some source of electrons.

The formula which gives the dependence of the average electroluminescence brightness on the voltage is [Eq. (33)]

$$\overline{B} = A \exp\left(-\frac{b_1}{\sqrt{V}}\right),$$

where the pre-exponential factor A represents an external source of electrons and the argument of the exponential function represents the field concentration conditions.

The energy absorbed by a phosphor will be considered on the basis of the ideas developed in the investigation of the voltage dependence of the efficiency, and the conditions will be assumed to be the same as those obtaining in that investigation. We shall postulate, as before, that the energy is absorbed mainly by the electrons liberated from the local levels by phonon-assisted tunnel transitions, whose probability is given by Eq. (41). The absorbed energy is then given by the expression

$$W = qV A_2 \exp b_2 V. \tag{48}$$

The quantity A_2 represents the strength of an internal source of electrons which participate in the energy absorption. This quantity is determined by the density of the levels from which electrons are liberated by the field, and the coefficient b_2 is determined by the depth of these levels.

We used our concept of the mechanism of electroluminescence to predict the dependences of the coefficients A, b_1 and A_2, b_2 on the properties of a phosphor, which were governed by the conditions during its preparation. Next, we tested the validity of our ideas by measuring these coefficients experimentally.

We investigated the same electroluminescent capacitors as in § 1. They were excited by an alternating voltage of line frequency at room temperature. The brightness and absorbed power were measured in the same way as in our study of the voltage dependence of the efficiency. It was necessary to carry out measurements on large batches of electroluminescent capacitors with different phosphors. The number of capacitors with the same phosphor varied from one batch to another and, in some cases, it reached 35 samples. We selected capacitors with approximately the same phosphor layer thickness d (about 0.06 mm) and introduced a correction for the thickness after measuring it carefully, i.e., we divided the value of b_1 by $\sqrt{d}$ and multiplied b_2 by d (making suitable allowance for the dimensions).

The first measurements were carried out on the PIAS and SIAC phosphors, which had approximately the same copper concentration ($\sim 10^{-3}$ g-atom/mole). The values of A, b_1 and A_2, b_2 obtained from the results for these capacitors are presented in Table 1. We found that the properties of all these samples were similar. The absorbed power was a few milliwatts at a voltage of 300 V and the maximum efficiency was 1.5-2%.

Next, we investigated a phosphor which also contained a second phase but had a much lower concentration of copper in the lattice. This phosphor was prepared from an ordinary photoluminescent material in which the concentration of Cu was 10^{-6} g-atom/mole. We treated the photoluminescent material in a solution of $CuSO_4$, which endowed it with electroluminescent properties. If the introduction of copper into the ZnS lattice had given rise to those levels which provided the energy-absorbing electrons, the value of A_2 should be less for an electroluminescent phosphor prepared in this way. Our experiments showed that the value of A_2 was indeed an order of magnitude smaller. Since it was known that our treatment produced a second phase and the phosphor become luminescent in an electric field, we concluded that this phosphor had a source of electrons which was probably weaker than the source in conventional electroluminescent phosphors. The field concentration conditions in the phosphor treated with copper sulfate should be much poorer (b_1 should be larger) if the copper gave rise to those centers at which space charge could accumulate. The experimental results confirmed this prediction (Table 1).

In a second experiment, we tried to retain the same conditions inside a crystal by comparing two samples with very different amounts of the second phase. This was done by taking two samples in which the concentration of copper was exactly the same (10^{-3} g-atom/mole) and subjecting them to the same treatment, except that one of them (the SIAC phosphor) was cooled

TABLE 1. Measured Values of the Parameters A, b_1, A_2, and b_2

Phosphor	A	b_1	A_2	b_2
Electroluminescent PIAS	1.43	354	$1.64 \cdot 10^{-3}$	$2.33 \cdot 10^{-4}$
Electroluminescent SIAC	3.06	286	$4.57 \cdot 10^{-3}$	$2.58 \cdot 10^{-4}$
Photoluminescent after treatment	0.19	657	$2.22 \cdot 10^{-4}$	$1.39 \cdot 10^{-4}$
Quenched	$3.04 \cdot 10^{-2}$	335	$1.26 \cdot 10^{-2}$	$1.43 \cdot 10^{-4}$
Photoluminescent before treatment	—	—	$5.40 \cdot 10^{-5}$	$2.33 \cdot 10^{-4}$

slowly to room temperature and the other was quenched by dropping it into liquid nitrogen.* Rapid cooling should prevent the precipitation of copper as a second phase, and its concentration in the lattice should be considerably higher, corresponding to the high-temperature equilibrium value. Such a phosphor should contain more levels of the kind from which energy-absorbing electrons could be liberated. It was found that the value of A_2 for this phosphor was higher than that for a phosphor cooled in the usual slow manner. The field concentration conditions in the quenched phosphor should have been approximately the same (or even somewhat better) in a normally cooled sample. The experimentally determined value of b_1 was found to be similar to (but somewhat larger than) the value for a normally cooled sample. However, it was not possible to distinguish these values of b_1 because the difference between them was less than the experimental error of 10%. The conditions used in the preparation of these phosphors suggested that the source of electrons in the quenched material should be weaker. The value of A for this phosphor was indeed two orders of magnitude smaller than that for a normally cooled phosphor.

We also measured the energy absorption and determined the parameters A_2 and b_2 for a conventional photoluminescent phosphor, in which the concentration of Cu was $\sim 10^{-6}$ g-atom/mole. The value of A_2 of this phosphor was the lowest of those tested: after the treatment of the phosphor in copper sulfate, the value of A_2 was slightly higher but still well below the values for the electroluminescent phosphors (especially the quenched sample). The difference between these values was much greater than the experimental error, which was 20% for the parameter A_2.

The last column of Table 1 gives the values of b_2 for the deepest levels from which electrons were liberated only by fields of critical or higher value. The values of b_2 for all the samples were equal within the limits of the experimental error, which was 30%. Therefore, we may assume that the deepest levels were the same in all the phosphors investigated. These results confirm that the liberation of electrons from the deep levels associated with activator atoms governs the energy absorbed by the phosphor.

On the other hand, the shallow levels play different roles in different phosphors. The differences between the shallow levels should be manifested particularly at low voltages because they are empty at high voltages. Figure 7 shows the dependences log W/V = f(V) for capacitors containing three different phosphors. We can see that the experimental points for the photoluminescent phosphor fit a straight line right down to low voltages, i.e., the shallow levels are not active in the process of energy absorption. For the quenched electroluminescent phosphor there is an appreciable deviation of the experimental points from the straight line in the low-voltage region. This deviation is even greater for untreated good-quality electroluminescent phosphors (curve 2 in Fig. 7), which means that the shallow levels in these phosphors do participate appreciably in the energy absorption.

These experiments give quantitative information on the separate influences of the field concentration conditions and of the source of electrons on the excitation of electroluminescence,

* The author is grateful to O. N. Kazankin for preparing these phosphors.

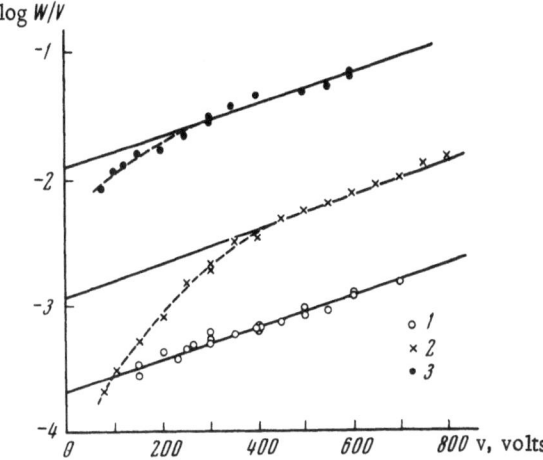

Fig. 7. Dependence of the absorbed power on the voltage for three electroluminescent capacitors: 1) capacitor containing a photoluminescent phosphor (the same results were obtained after heat treatment of this phosphor); 2) capacitor containing an electroluminescent phosphor; 3) capacitor containing a quenched phosphor.

as well as on the role of the deep levels in the absorption of energy by a phosphor. Moreover, the results obtained help one to determine the importance of the various factors which govern the electroluminescence efficiency.

The amount of energy absorbed in a phosphor decreases when the concentration of copper is reduced. However, this does not result in an increase of the efficiency because of the simultaneous (and stronger) decrease in the brightness. It follows from our discussion that the quantities b_1 and A_2, associated with field concentration conditions and the value of the absorbed energy, may be related to each other because both are governed by the density of the electron traps in a crystal. However, it may happen that the traps which govern the field concentration conditions are different from those which determine the absorption of energy. Then, the parameters b_1 and A_2 may be independent. The existence of the postulated relationship has not yet been established.

It is clear from our discussion that the levels which determine the energy absorption should be shallow because such a condition would reduce the amount of energy which goes into the nonradiative channel. If, moreover, these shallow levels are of the donor type, i.e., if they retain the positive space charge necessary for the field concentration in the excited state, the brightness will be stronger and, therefore, the efficiency will be higher. Naturally, the efficiency cannot exceed its theoretical limit [28] , which is about 10% for the excitation mechanisms which occur in ZnS:Cu phosphors.

CONCLUSIONS

The theoretical discussion of the kinetics of the Destriau effect given in the present paper represents a development (in the quantitative sense) of the ideas advanced by A. N. Georgobiani and M. V. Fok. However, the transport equations are based on very general assumptions about the band structure of a crystal phosphor with three levels in the forbidden band. No *a priori* assumptions are made about the excitation mechanism. Therefore, these general transport equations can be used to determine which of the possible excitation mechanisms does indeed occur in a given crystal phosphor. Thus, they are suitable for the investigation of various substances.

A considerable simplification of the general system of transport equations is achieved by making only three fairly general and experimentally justified assumptions about the Destriau effect:

1. The electroluminescence process is assumed to occur in two steps or stages;

2. A high-field region (field concentration) is assumed to be established in a crystal phosphor grain during its excitation;

3. It is assumed that the charges drift in an electric field in the interior of a crystal.

Further simplification of the band scheme is based on specific assumptions about the nature of the space charge in the high-field region and, therefore, it is valid only for those substances for which these assumptions are justified. The system of equations obtained in this way can be solved by means of quadrature formulas. If the electric-field dependences of the various quantities occurring in these equations are known explicitly from independent experiments, the solution can be obtained by numerical calculations.

The system of transport equations can be used to refine the physical ideas on the Destriau effect and to determine their internal self-consistency, which is not possible when the individual characteristics are considered separately. By way of example, this approach is applied to the characteristic features of the establishment of an excited state in a crystal.

The theory is used to derive the dependences of the brightness and the absorbed power on the applied voltage. This derivation is based on the assumption that the excitation of a crystal occurs in two stages: first, electrons provided by some external source penetrate the high-field region where they are accelerated to cause impact ionization of the crystal lattice and of the luminescence centers. It is assumed that the energy is absorbed by the electrons liberated by the ionization of the crystal lattice and of the luminescence centers, as well as by the electrons liberated by the field from the deep levels (the second absorption process is the dominant one).

The theory is compared with experiment by measuring the energy absorbed from an alternating electric field. The dependence of the electroluminescence efficiency on the applied voltage is determined for capacitors containing various types of ZnS:Cu phosphor. In these experiments, the two theoretical parameters b_1 and b_2 are deduced from independent data. The coefficient b_2, representing the depth of the donor levels, is found by three different methods. All three methods give values which agree quite satisfactorily for the same phosphor. The differences between the parameters for different phosphors are in agreement with the theoretical predictions.

Thus, the expression for the electroluminescence efficiency derived from the transport equations is confirmed experimentally. This means that the ideas on the kinetics of the Destriau effect deduced from the investigation of the emitted electroluminescence are also supported by studies of the absorbed energy.

The author is deeply grateful to M. V. Fok for his continuous interest and great help in this investigation.

LITERATURE CITED

1. G. Destriau, J. Chim. Phys., 33: 620 (1936).
2. H. K. Henisch, Electroluminescence, Pergamon Press, Oxford (1962).

3. A. N. Georgobiani, Tr. Fiz. Inst. Akad. Nauk SSSR, 23: 3 (1963) [Soviet Researches on Luminescence, Consultants Bureau, New York (1964), p. 1].
4. Yu. P. Chukova, Tr. Fiz. Inst. Akad. Nauk SSSR, 37: 149 (1966) [Electrical and Optical Properties of Semiconductors, Consultants Bureau, New York (1968), p. 127].
5. V. M. Fok, Introduction to the Kinetics of Luminescence of Crystal Phosphors [in Russian], Nauka, Moscow (1964).
6. K. S. Rebane and É. K. Tal'viste, Tr. Inst. Fiz. Astron. Akad. Nauk Est. SSR, 15: 161 (1961).
7. K. S. Rebane, Opt. Spektrosk., 12: 396 (1962).
8. M. V. Fok, Opt. Spektrosk., 11: 98 (1961).
9. E. Yu. L'vova, Opt. Spektrosk., 22: 600 (1967).
10. E. E. Bukke, L. A. Vinokurov, and M. V. Fok, Opt. Spektrosk., 16: 491 (1964).
11. A. N. Georgobiani, Opt. Spektrosk., 11: 426 (1961).
12. A. N. Georgobiani and M. V. Fok, Opt. Spektrosk., 11: 93 (1961).
13. A. N. Georgobiani and M. V. Fok, Opt. Spektrosk., 5: 167 (1958).
14. A. N. Georgobiani and M. V. Fok, Opt. Spektrosk., 9: 775 (1960).
15. A. N. Georgobiani and M. V. Fok, Opt. Spektrosk., 10: 188 (1961).
16. A. N. Georgobiani and Yu. G. Penzin, Opt. Spektrosk., Sbornik I, Luminestsentsiya, p. 321 (1963).
17. A. N. Georgobiani, E. G. Matinyan and A. N. Savin, Opt. Spektrosk., 18: 347 (1965).
18. M. V. Fok, Usp. Fiz. Nauk, 72: 467 (1960).
19. V. A. Chuenkov, Fiz. Tverd. Tela, Sbornik Vol. 2, 200 (1959).
20. L. V. Keldysh, Zh. Eksp. Teor. Fiz., 37: 713 (1959).
21. C. M. Zener, Proc. Roy. Soc., London A145: 523 (1934).
22. I. K. Vereshchagin and I. T. Drapak, Opt. Spektrosk., Sbornik I, Luminestsentsiya, p. 327 (1963).
23. E. E. Bukke, L. A. Vinokurov, and M. V. Fok, Inzh.-Fiz. Zh., 1(7):113 (1958).
24. L. V. Keldysh, Zh. Eksp. Teor. Fiz., 34: 962 (1958).
25. A. N. Georgobiani, E. Yu. L'vova, and M. V. Fok, Opt. Spektrosk., 13: 564 (1962).
26. A. N. Georgobiani, E. Yu. L'vova, and M. V. Fok, Opt. Spektrosk., 15: 95 (1963).
27. A. N. Georgobiani, E. Yu. L'vova, and M. V. Fok, Opt. Spektrosk., 15: 266 (1963).
28. M. V. Fok, Opt. Spektrosk., 18: 1024 (1965).

ELECTROLUMINESCENCE AND OTHER OPTICAL AND ELECTRICAL CHARACTERISTICS OF p—n JUNCTIONS IN ZINC SULFIDE

A. N. Georgobiani and V. I. Steblin

A method for the fabrication of p—n junctions in zinc sulfide was developed. The current—voltage characteristics, phto-emf, and electroluminescence of these junctions were investigated. The results obtained were used to estimate the barrier height, donor level depth, and forbidden band width of ZnS single crystals. The mechanism of current flow through a p—n junction was analyzed. Information on the electron energy states and the mechanism of electroluminescence was obtained from the electroluminescence characteristic. A comparison was made between the characteristics of p—n junctions in zinc sulfide and those of $ZnS:Cu_2S$ heterojunctions. This comparison was used to obtained more information on the mechanism of the electroluminescence of ZnS single crystals.

INTRODUCTION

The rapid development of semiconductor technology has stimulated efforts to extend and deepen our knowledge of the structure of solids and thus has made it possible to obtain very valuable information on the physical processes in solids.

The success in the fabrication of semiconductor devices has been, to a considerable extent, due to the considerable improvements in the preparation of pure and very pure substances. The concentration of impurities in high-quality semiconducting materials is usually $< 10^{-8}\%$, which represents less than one impurity atom per 10^{12} host atoms. However, some semiconductor devices can be prepared from substances of lower purity. For example, materials with high impurity concentrations (degenerate semiconductors) are used currently in the fabrication of lasers, tunnel diodes, and other devices.

There are many semiconducting compounds which have as yet no applications in technology but exhibit some interesting and technically valuable properties. The useful application is prevented by the serious difficulties encountered in the preparation and purification of some of these compounds and the consequent difficulties in controlling their physical properties.

The continuous development of science and technology is extending the range of requirements which semiconductor devices are expected to meet. Materials stable at high temperatures are needed for high-power devices and for circuits in which high current overloads may occur. Electronic apparatus is required for operation at ambient temperatures of 300-500°C, or at even higher temperatures. The germanium and silicon diodes and transistors manufactured by the Soviet semiconductor industry are unsuitable for this purpose. Germanium devices can be operated at ambient temperatures up to 100°C. The working temperatures of silicon devices are somewhat higher (up to 180°C). Rectifiers made of TiO_2 and GaAs are capable of working at temperatures of 200-300°C. Silicon carbide seems to be the only currently available material which can be used in high-temperature rectifiers. These rectifiers have very good characteristics even at ambient temperatures exceeding 500°C.

The ambient temperature at which a given semiconductor device can operate is determined by the thermal stability and forbidden band width of the material used in that device. The wider the forbidden band of a semiconductor, the higher is the temperature at which a device made of it can operate. This is in good agreement with the examples just quoted: germanium has the narrowest forbidden band (0.7 eV) and SiC the widest (2.2-3.2 eV, depending on the modification) among the materials mentioned in the preceding paragraph. Semiconducting materials with wide forbidden bands are also needed in short-wavelength radiation detectors, noncoherent light sources, and lasers operating in the visible or ultraviolet parts of the spectrum.

The emission of light in semiconductors may result from the recombination of free electrons and holes (interband recombination) or of free carriers and ionized luminescence centers; or it may result from the transitions from the excited to ground states in such centers (intracenter luminescence). On the short-wavelength side, the luminescence spectrum of a semiconductor is usually limited by its forbidden band width. Therefore, visible luminescence can be obtained provided the forbidden band width is ≥ 1.8 eV and ultraviolet radiation can be obtained only if this band width is ≥ 3.1 eV. The intensity of the luminescence is governed by the number of radiative recombination events. Consequently, a semiconducting material suitable for the fabrication of high-efficiency electroluminescent light sources should exhibit a high probability of radiative recombination compared with the probability of nonradiative processes. Moreover, the efficiency of a semiconductor light source depends on the method used to transfer electrons from the valence to the conduction band or to excite luminescence centers. There are three main mechanisms of excitation by an external electric field:

(1) injection of minority carriers (for example, through a p − n junction biased in the forward direction);

(2) tunneling (Zener effect), in which electrons are liberated by an electric field from the levels of the luminescence centers, or from the valence band, and are then transferred to the conduction band;

(3) impact ionization: in this case, electrons (or holes) are accelerated by a strong electric field and thus acquire an energy sufficient for the ionization or excitation of luminescence centers or for the transfer of electrons from the valence to the conduction band.

An analysis of the radiant efficiency of electroluminescence excited by these various methods was made by M. V. Fok [1]. He showed that an electroluminescent source of light had the highest efficiency if it was excited by injection.

$A^{II} B^{VI}$ compounds, used widely in the preparation of phosphors, are worth serious consideration because of their wide forbidden bands and their high probabilities of radiative recombination. The most extensively used are phosphors based on zinc sulfide (thermal forbidden band width $E_T = 3.2 + 0.2$ eV [2] and optical width $E_0 \approx 3.7$ eV [3]). Fischer [4] calculated the intensity of luminescence emitted by an ideal p−n junction in zinc sulfide carrying a forward current of 1 A/cm^2 density (studies of cathodoluminescence demonstrated that this substance was capable of withstanding such high excitation levels). According to Fischer, the intensity of the green luminescence emitted by such a p−n junction should be 675 lm/cm^2.

It is difficult to fabricate p–n junctions in wide-gap materials (including zinc sulfide) because the achievement of n- and p-type conduction becomes progressively more difficult as the forbidden band gets wider. Zinc sulfide is known to be a high-resistivity n-type semiconductor. There is no published information on the preparation of p-type ZnS. The high resistivity of ZnS is attributed by many workers to deep impurity levels and to a charge-compensation mechanism resulting in electrical neutrality of the activator impurities. The predominance of n-type conduction in ZnS may be associated with the stronger volatility of sulfur, compared with that of zinc, during preparation. The excess zinc retained in the lattice gives rise to donor levels and, therefore, ZnS has n-type conduction.

The present paper describes the fabrication of p–n junctions in ZnS and a study of the optical and electrical characteristics of such junctions.

CHAPTER I

SOME PROPERTIES OF p–n JUNCTIONS

§1. Current–Voltage Characteristics in Darkness

The foundations of the modern theory of p–n junctions were laid in 1949 by Shockley [5]. One of the important assumptions made by Shockley is that the width of the space–charge region is small compared with the diffusion length. Therefore, the generation of carriers in the space–charge region can be neglected. The Shockley current − voltage characteristic is given by

$$\mathcal{J} = \mathcal{J}_s\left(\exp\frac{eV}{kT} - 1\right), \tag{1}$$

$$\mathcal{J}_s = \frac{eD_p p_n}{L_p} + \frac{eD_n n_p}{L_n}, \tag{2}$$

where $\mathcal{J}_s$ is the saturation current; $\mathcal{J}$ is the current through the p–n junction; e is the electronic charge; V is the applied voltage; k is the Boltzmann constant; T is the absolute temperature; D_n, D_p are, respectively, the diffusion coefficients of electrons and holes; p_n is the density of holes in the n-type region; n_p is the density of electrons in the p-type region; L_n is the diffusion length of electrons in the n-type region; and L_p is the diffusion length of holes in the p-type region.

The characteristic features of this relationship are:

(a) saturation of the current at reverse bias values corresponding to a few kT/e;
(b) an exponential rise of the current with increasing forward voltage (this voltage is assumed to be concentrated in the p–n junction).

Investigations of germanium p–n junctions confirmed the validity of the Shockley theory. The only experimental observation which required an additional explanation was the slow rise of the reverse current with increasing voltage (proportional to $V^{1/2}$ or $V^{1/3}$) in the region where the Shockley theory predicted saturation.

The thermal generation and the recombination of carriers in the space–charge region of a p–n junction can be neglected only at low values of the reverse voltage. Such generation and recombination are allowed for in the treatment due to Tolpygo and Rashba [6]. They thought it necessary to obtain the solution for the whole of the semiconductor and not only outside the space–charge region, as was done by Shockley. Tolpygo and Rashba considered direct exchange between the allowed bands as the generation and recombination mechanism.

They assumed that the contribution of the processes involving traps was negligible. In this theory, electrons are transferred by thermal fluctuations from the valence to the conduction band. The reverse process of carrier recombination takes place simultaneously with the process of generation. Under equilibrium conditions the number of transitions in both directions are equal. When a reverse voltage is applied to the junction, the carriers move from the junction to the electrodes. Therefore, those carriers which are generated in the space-charge layer are prevented from recombining by their drift in the applied field. Consequently, the rate of thermal generation of carriers in the p−n junction region becomes greater than the rate of recombination. This gives rise to an additional component of the current [6], whose value is directly proportional to the width of the space-charge region. The saturation current now becomes

$$\mathcal{I}_s = e \left(\frac{X_1}{\tau_p} + \frac{D_p}{L_p} \right) p_n + e \left(\frac{X_2}{\tau_n} + \frac{D_n}{L_n} \right) n_p, \tag{3}$$

where $X_{1,2}$ are the dimension of the space-charge layer in the n- and p-type regions; τ_n and τ_p are the electron and hole lifetimes, respectively.

The saturation thus depends on the width of the space-charge layer and, therefore, on the applied voltage. Tolpygo and Rashba [6] considered abrupt ($\mathcal{I}_s \propto V^{1/2}$) and linear-gradient ($\mathcal{I}_s \propto V^{1/3}$) junctions. Experiments carried out by Kosenko [7] on germanium diodes confirmed the correctness of Tolpygo and Rashba's basic assumptions. It was also found that the Shockley theory was unable to account for many properties of p−n junctions in silicon. The current−voltage characteristics of such junctions are described by the empirical formula

$$\mathcal{I} = \mathcal{I}_s \left(\exp \frac{eV}{AkT} - 1 \right). \tag{4}$$

The temperature dependence of the reverse current is considerably weaker than that predicted by the Shockley theory [5]. Moreover, the argument of the exponential function describing the forward current includes the ratio e/AkT, where A > 1. We now know that similar properties are exhibited also by p−n junctions made of other materials with wider gaps (GaAs, InP, SiC, CdS, CdTe, and ZnTe).

Sah, Noyce and Shockley [8] explained the behavior of silicon diodes by making allowance for carrier generation and recombination in the space-charge region of the junction. According to their treatment, carrier pairs are generated in the space-charge layer when a reverse voltage is applied to a p−n junction. These pairs give rise to a generation current

$$\mathcal{I}_g = - We \frac{n_i}{2\tau_0}, \tag{5}$$

where n_i is the intrinsic carrier density; τ_0 is the carrier lifetime in an intrinsic semiconductor; W is the width of the space-charge layer.

The current due to carrier generation in the neutral regions (the diffusion current) is given by the expression

$$\mathcal{I}_d = - 2e \frac{D}{L} \frac{n_i^2}{n_n}. \tag{6}$$

A comparison of these two expressions shows that the current representing the carriers generated in the space-charge layer may exceed considerably the diffusion current in semiconductors with short lifetimes, low resistivities, and wide gaps, even when the dimensions of the space-charge layer are smaller than the diffusion length.

Under forward voltages, the dominant process in the space-charge layer is carrier recombination. The recombination current is given by the expression

$$\mathcal{J}_r = 2 \left(\frac{kT}{eE}\right) \frac{n_i}{2\tau_0} \exp \frac{eV}{2kT} , \qquad (7)$$

where E is the field in the junction. The total current through the junction is equal to the sum of the diffusion current of the Shockley theory [5] and the generation – recombination current of the Sah–Noyce–Shockley theory [8]:

$$\mathcal{J} = \mathcal{J}_r + \mathcal{J}_d. \qquad (8)$$

The assumption that recombination predominates under forward voltage conditions gives rise to a dependence of the forward current on the voltage, which is of the type $\exp(eV/AkT)$, where A = 2. This dependence explains the typical features of the forward current–voltage characteristics of p–n junctions in various wide-gap materials.

Sah, Noyce, and Shockley tested their theory experimentally [8]. They investigated diffused silicon diodes having special geometry which eliminated the influence of surface leakage. This test demonstrated that the theory agreed with the experiment over a range of currents exceeding nine orders of magnitude. Belova, Kovalev, and Penin [9] discovered that carrier generation in the space-charge layer of a germanium p–n junction had a considerable influence on the reverse current-voltage characteristic of such a junction. They found that the results were in qualitative agreement with the Sah–Noyce–Shockley theory. Much poorer agreement between the experimental results and the Sah–Noyce–Shockley theory was obtained by Nasledov and Tsarenkov for diffused p–n junctions in GaAs]10]. Their forward current-voltage characteristic was given by the expression

$$\mathcal{J} = \mathcal{J}_0 \exp \frac{eV}{AkT} ,$$

where A > 2. The pre-exponential factor was three-five orders of magnitude greater than the Sah–Noyce–Shockley recombination current. The reverse current did not vary in accordance with the $V^{1/3}$ law but much more rapidly. Similar characteristics were reported for alloyed p–n junctions in GaAs [11]. Much later investigations of p–n junctions in GaAs [12] demonstrated that the coefficient A was a function of the applied voltage and of the recombination center parameters.

We have considered so far a p–n junction, ignoring the bulk resistances of the p- and n-type regions. This approach is valid only if these resistances are very low. If one of the regions has a high resistivity, the carriers injected through the junction may modulate considerably the resistance of this region. This effect is observed at high injection levels. The influence of the bulk resistance of a diode on its characteristics was first investigated by Stafeev [13].

§2. Electroluminescence

In the preceding section, we have considered the influence of carrier generation and recombination in the p–n junction region on the current–voltage characteristic. Since recombination may be radiative or nonradiative, it is natural to assume that the application of a forward voltage should give rise to some emission of radiation by the p–n junction. Such radiation is emitted also by a reverse biased junction but the excitation mechanism differs basically from the mechanism dominant in the case of forward voltages. We shall consider these two mechanisms separately.

A. Luminescence of Forward Biased p − n Junctions. Systematic investigations of the luminescence emitted by forward biased p−n junctions were started by Newman [14], who studied germanium. Subsequently, injection electroluminescence was intensively studied in Si[15], SiC[16], GaAs [17], GaP[18], ZnTe[19], ZnSe[20] and other components.

The luminescence spectra of germanium p−n junctions are described in Newman's papers [14, 21]. Newman established the existence of an intrinsic luminescence band and a band with a maximum at about 0.5 eV, which was independent of the impurity concentration. This long-wavelength band was attributed to dislocations [21]. A systematic study of the influence of structure defects on recombination radiation emitted by germanium was carried out by Gippius and Vavilov [22].

Observations carried out on Si confirmed the importance of the interband recombination [15]. Apart from this interband luminescence, the spectrum of silicon included also additional peaks which were sensitive to the presence of impurities and were therefore attributed to a recombination process involving the participation of impurities.

The luminescence spectra of p−n junctions in GaAs, GaP, and SiC are systems of bands representing the interband recombination of injected electrons and holes, as well as recombination at impurities.

At liquid nitrogen temperature, the luminescence spectrum of p−n junctions in ZnTe [19], fabricated by alloying with indium, consists of two bands: a green band, about 6 nm wide, with a maximum near 540 nm, and a red band extending from 580 to 700 nm. This spectrum is shown in Fig. 1a. The relative intensities of these two bands vary with temperature: at higher temperatures, the red band predominates; at lower temperatures, the green band is stronger. A comparison with the photoluminescence spectrum [23] demonstrates that the green band is due to the recombination of free electrons or of electrons captured by shallow traps with holes located at the acceptor levels of phosphorus. The red band may be associated with indium impurities and may be attributed to the recombination of electrons localized at deep donor levels with holes from the valence band or holes trapped at shallow acceptor levels.

ZnSe diodes [2] luminesce at shorter wavelengths than ZnTe junctions. The injection electroluminescence spectrum of a zinc selenide diode, recorded at 130°K, is shown in Fig. 2. The short-wavelength tail, extending to 2.58 eV, is attributed in [20] to interband transitions. This energy is close to the absorption edge of pure ZnSe, which is located at 2.68 eV [24]. Thus,

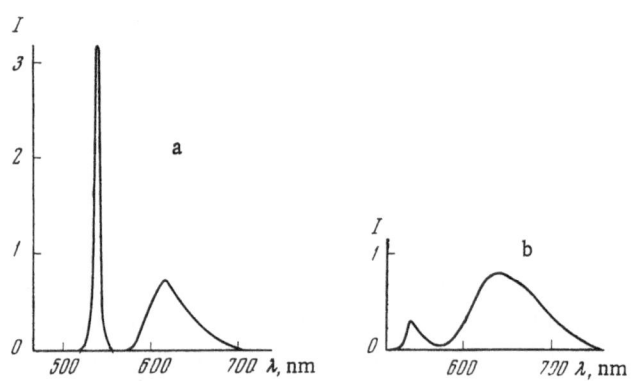

Fig. 1. Injection electroluminescence spectrum of a ZnTe diode at liquid nitrogen temperature (a) and the luminescence spectrum of the same diode under avalanche breakdown conditions (b).

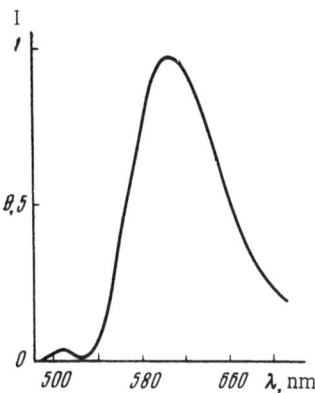

Fig. 2. Electroluminescence spectrum of a forward biased p–n junction in ZnSe.

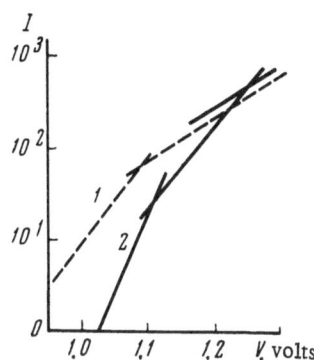

Fig. 3. Dependence of the brightness of the luminescence emitted by GaAs diode on the applied voltage. 1) Band with a maximum at 0.97 eV; 2) band at 1.21 eV.

a characteristic feature of forward biased p–n junctions in zinc selenide is the presence of luminescence bands representing the recombination of conduction-band electrons with valence-band holes, as well as the recombination associated with levels due to impurities or structure defects.

The dependence of the electroluminescence brightness on the forward current through a p–n junction is complex: it is governed by the method used in the fabrication of this junction, the temperature at which the measurements are carried out, the voltage used, and other parameters. At voltages exceeding the p–n junction barrier height, this dependence is usually linear. Such behavior has been reported for germanium [25], silicon [26], and other diodes mentioned in the preceding paragraphs. The brightness is proportional to the current when the following two conditions are satisfied [27]:

  (a) the injection efficiency is constant (independent of the current);
  (b) the ratio of the rates of radiative and nonradiative recombination is independent of the injection level.

Figure 3 shows the dependence of the brightness of the luminescence emitted by a GaAs diode on the applied voltage [12]. Similar dependences have been reported for Si [26], ZnTe [19], and ZnSe [20]. The dependence presented in Fig. 3 was explained by Morgan [28]. He considered theoretically the recombination processes occurring at deep levels and took into account the influence of a forward voltage on the Fermi level position in the junction region. His results were in good agreement with the experimental data reported in [12] for GaAs.

B. Luminescence during Breakdown of p – n Junctions. The luminescence of Si diodes under breakdown conditions was first reported by Newman in 1955 [29]. Similar breakdown luminescence was later reported for Ge [30], GaP [31], GaAs [32], SiC [33], ZnTe [23], and ZnSe [20].

All p–n junctions exhibiting breakdown luminescence can be divided into two types: the avalanche junction, in which the breakdown mechanism is of the avalanche type; and the tunnel junction, in which the breakdown is of the Zener type. The breakdown mechanism determines the quantitative and qualitative characteristics of the luminescence. Under breakdown conditions, the light is usually emitted by small isolated microplasma regions. The luminescence is observed in that part of the characteristic in which the current rises rapidly. The number of luminous microplasma regions (points) grows with increasing breakdown current, i.e., the total luminescence yield increases with the current. It has been recently established that the luminous micro-plasmas are concentrated at dislocations which penetrate the junction. A reduction of the length of the dislocations by improvement of the technology of preparation gives a more or less uniform emission over the whole junction [34]. Table 1 lists some of the characteristics of the luminescence emitted by seven semiconductors ranked in the order of increasing forbidden band width. The information is fullest in the case of silicon and germanium diodes. Only silicon p−n junctions exhibit luminescence under Zener and avalanche breakdown conditions. No information is available on the luminescence spectrum under Zener conditions [36]; the avalanche breakdown luminescence [38] is proportional to the current and its spectrum consists of a very wide band. Avalanche breakdown is reported in [41] for GaAs but no data are given on the luminescence. Kholuyanov [33] observed blue and white microplasma spots during breakdown in SiC diodes. The luminescence spectrum of these diodes was a broad band little affected by the current flowing through the junction.

Figure 4 shows the room temperature luminescence spectra of Ge, Si, and SiC diodes. At high energies, the intensity in these spectra decreases monotonically to the edges occurring at 2.1, 3.1, and 5.9 eV for Ge, Si, and SiC, respectively. In the long-wavelength range, these spectra behave differently [33, 35, 36]. Chynoweth and McKay [36] have suggested that the short-wave-length luminescence of Si p−n junctions is associated with phonon-assisted recombination of free electrons and holes in the breakdown plasma. Theoretically, a photon of 5.7 eV energy may result from the recombination of "hot" electrons and holes whose energies are close to the ionization threshold [37]. However, the probability of the creation of such a photon is extremely low. In the case of the recombination of a "hot" electron and a thermal hole, or conversely, the maximum energy of photons cannot exceed 3.4 eV. This value is in agreement with the observed edge at 3.1 eV. A similar analysis can be carried out for SiC and Ge. The low-energy end of the luminescence spectrum is attributed in [38, 39] to transitions within the allowed band (arrow 5 in Fig. 5). In this intraband mechanism, the maximum photon energy cannot exceed the ionization energy E_i. However, if such transitions do occur, there should be some local change in the spectrum close to E_i. It is likely that the peak at 1.18/eV in the spectrum of Ge is due to this cause (E_i = 1.5 eV). However, there are no corresponding changes in the spectra of Si (E_i = 2.3 eV) and SiC (E_i = 4 eV).

TABLE 1

Semicon-ductor	Breakdown mechanism	Lumi-nescence	Nature of spectrum	Reference
Ge	Avalanche Zener	+ —	Structured	[30,35]
Si	Avalanche Zener	+ +	Smooth No information	[38] [36]
GaAs	Avalanche Zener	+ —	Structured	[41]
ZnTe	Avalanche Zener	+ —	Structured	[23]
GaP	Avalanche Zener	+ —	No information	[31]
SiC	Avalanche Zener	+ —	Smooth	[33]
ZnSe	Avalanche Zener	+ —	Structured	[20]

*The plus sign indicates that luminescence is observed and the minus sign that it is not.

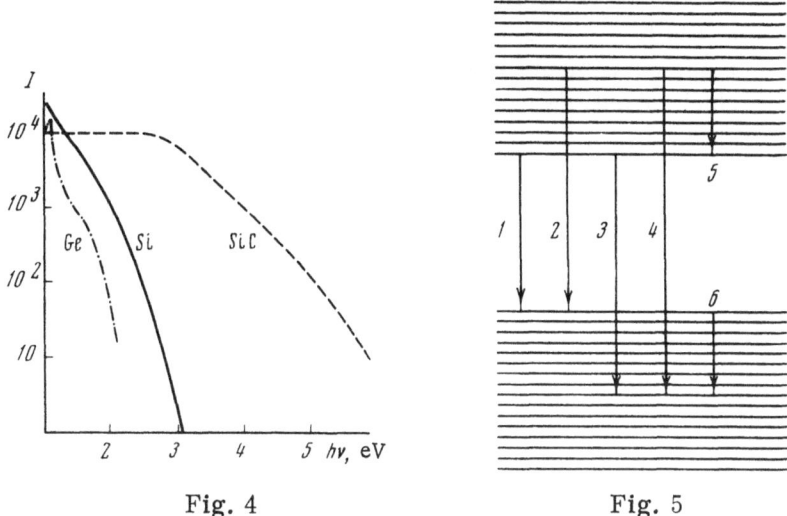

Fig. 4   Fig. 5

Fig. 4. Breakdown luminescence spectra of p—n junctions in Ge, Si, and SiC.

Fig. 5. Schematic representation of electron transitions during the breakdown of p—n junctions in Ge, Si, and SiC: 1) interband recombination of electrons and holes; 2) recombination of a "hot" electron and a hole; 3) recombination of a "hot" hole and an electron; 4) recombination of a "hot" electron and a "hot" hole; 5), 6) intraband transitions.

In contrast to the three compounds discussed in the preceding paragraph, the breakdown luminescence spectrum of a ZnTe diode (Fig. 1b) is of the same nature as the spectrum obtained under a forward bias (Fig. 1). However, there are some differences between the two spectra: under a reverse bias, the ratio of the intensities of the green and red bands is less than that for a forward biased diode, the red band is wider, and luminescence between the green and the red bands is observed when the current is increased. The similarity of the luminescence spectra obtained under forward and reverse bias conditions indicates that the same luminescence centers are excited in both cases. This has been observed also for GaAs diodes [40]. Watanabe [19] attributes the luminescence emitted by reverse-biased ZnTe diodes to the ionization of the luminescence centers in an avalanche breakdown. The absence of radiation corresponding to the interband recombination of electrons and holes is probably due to an insufficiently high value of the current through these diodes.

§3. Photo-Emf

In the preceding section, we have considered the luminescence emitted by a biased p—n junction. The converse effect, a photo-emf, appears when the junction is illuminated with light of wavelengths lying within an absorption band. We shall now consider the mechanism of the appearance of a photo-emf by assuming that strongly absorbed light is incident on the p-type region of a diode. Electron—hole pairs generated in this region by the incident light are separated when they approach the junction. The holes remain in the p-type region and the electrons are driven by the junction field across the barrier layer into the n-type region. This separation process occurs only if an electron—hole pair is generated at a distance from the p—n junction which is less than the diffusion length. In this way, minority carriers cross the junction and accumulate in one region, whereas majority carriers accumulate in the other region. This increase in the hole density in the p-type region and in the electron density in the n-type region is accompanied by an increase in the electric field established by these carriers. This field impedes the minority carrier transitions from one region to the other across the junction. The reverse current increases with increasing value of this field. The final result is a dynamic equilibrium when the currents of the minority carriers crossing the junction balance each other out. When this happens, a certain potential difference is established between the electrodes and this difference is, in fact, the photo-emf.

A photo-emf appears only when the energy of the incident photons $h\nu$ is greater than or equal to the forbidden band width of the semiconductor. However, if there are impurity levels in the forbidden band, a photo-emf may appear even when the photon energy is $h\nu < E_0$, where E_0 is the forbidden band width. A typical photo-emf spectrum of a p−n junction in ZnSe is shown in Fig. 6 [42]. It is evident from this figure that the photo-emf decreases with increasing photon energy in the range $h\nu > E_0$. This is probably due to weak penetration of photons of these energies into the semiconductor. When the illumination intensity is increased, the photo-emf increases as long as the carrier separation mechanism is operative, i.e., as long as a barrier still exists in the junction region. Thus, the maximum value of the photo-emf of a p−n junction cannot exceed the height of the barrier at this junction. Consequently,

$$V_{max} = \frac{F_p - F_n}{e} \, ,$$

where V_{max} is the maximum value of the photo-emf; F_p and F_n are the Fermi levels in the p- and n-type regions, respectively. Thus, the maximum value V_{max} which can be obtained for any given semiconductor should not exceed its forbidden band width. However, it is known that zinc sulfide [43] and substances such as PbS [44-47], ZnSe [48], and CdTe [49-50] exhibit photo-emfs exceeding considerably the forbidden band widths of these semiconductors. In the case of ZnS and ZnSe, the photo-emf may reach a few hundreds of volts. In the case of PbS and CdTe, this effect is attributed to the presence of p−n junctions which are distributed in such a way that their individual emf's are apt to produce a large photo-emf. One such possible configuration is shown in Fig. 7 [51]. The high values of the photo-emf's exhibited by ZnS and ZnSe are attributed to stacking faults in the crystal lattice. They may be regarded as alternating regions of cubic and hexagonal symmetries. It is assumed that each such defect is associated with a dipole layer and that all the layers have the same polarity.

§4. Fabrication of p − n Junctions

A. Diffusion Method. In this method, donor or acceptor impurities diffuse at elevated temperatures into the surface layer of a semiconductor. This produces a p−n junction between the bulk of the semiconductor and the thin surface layer containing the diffused impurity. During the diffusion, the semiconductor is placed in a furnace containing an atmosphere of the impurity vapor. The higher the temperature, the more rapid is the penetration of the impurity atoms into the semiconductor.

The depth of penetration thus depends on the temperature and the duration of the diffusion process. This temperature and duration are determined on the basis of the diffusion coefficient, the required concentration of the diffusant, and the desired thickness of the diffused layer.

Impurities can be diffused into a semiconductor not only from the vapor phase but also from liquids or solids. The deposition of a thin layer (of the order of a few tenths of a micron) of a substance containing an impurity which gives rise to an opposite type of conduction is followed by the application of a sufficiently high temperature. This causes the impurity to diffuse into the semiconductor to a predetermined depth at which a p−n junction is formed.

B. Alloying Method. This method is used most widely in the fabrication of p−n junctions. By way of example, we may consider alloying n-type germanium with indium, which

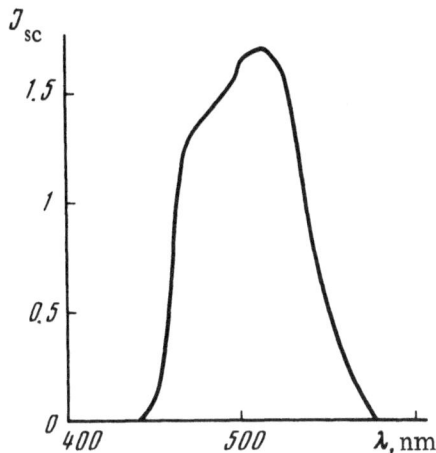

Fig. 6. Dependence of the short-circuit current on the wavelength of radiation incident on a ZnSe diode.

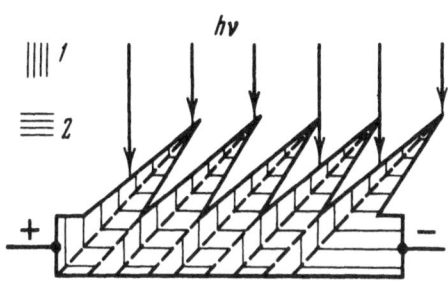

Fig. 7. Distribution of p–n junctions bringing about the summation of photo-emf's: 1) p-type; 2) n-type.

has acceptor properties in this semiconductor. A small pellet of indium is placed on a germanium slab and heated in vacuum or in a hydrogen atmosphere. The indium melts and wets the surface of the germanium. A surface layer of germanium dissolves in the indium, forming a small "pool" filled with a saturated solution of germanium in indium. If the temperature is lowered, the germanium crystallizes from the melt and the main part of the slab acts as a seed. Indium atoms are captured by the germanium during this recrystallization process and a surface layer with an excess concentration of acceptors is formed in this way. The concentration of indium increases away from the boundary between the recrystallized layer and the rest of the semiconductor. Outside the slab, the indium remains in the pure metallic form and is used to form an ohmic contact with the alloyed p-type Ge layer.

Thus, a p–n junction appears at the boundary between the recrystallized region and the rest of the slab. A pellet of lead containing some antimony is used to form a p–n junction on a p-type germanium slab. Aluminum is the most popular acceptor impurity used in the alloying of n-type silicon.

C. Heat Treatment Method. Rapid cooling of a semiconductor may sometimes cause inversion of its type of conduction. Such an inversion is due to the formation of defects at a high temperature and their freezing by rapid cooling. If a sharp metal point is pressed against a slab of n-type germanium and a high-current pulse is passed through this point, the region under the point is heated to a high temperature and then cools rapidly. This produces a p-type region and a p–n junction.

D. Introduction of Impurities into the Melt during Crystal Growth. We shall illustrate this method by considering germanium. The original material is purified and then melted in a crucible filled with an inert gas. A small single-crystal seed is lowered into the melt and then slowly pulled upward. An n-type germanium crystal solidifies on the seed in a form which depends on the temperature gradient at the solid-liquid boundary and on the rate of cooling. At some selected moment, an acceptor impurity is introduced into the melt and the amount of this impurity is governed by the desired acceptor concentration. Such doping of the melt produces a p-type crystal or region.

CHAPTER II

ELECTRICAL AND OPTICAL PROPERTIES OF ZINC SULFIDE

§1. Electrical Properties

Zinc sulfide occurs in nature in two forms: as zinc blende (cubic structure) and as wurtzite (hexagonal structure). The two modifications are very similar: the main difference is the relative distribution of the atomic planes in the lattice. Natural crystals of zinc sulfide are unsuitable for systematic investigations because they contain many impurities and it is difficult to estimate the concentration of these impurities. Therefore, investigations are usually carried out on single crystals of ZnS grown in the laboratory.

The electrical conductivity of zinc sulfide is usually quite low. It is due to donor levels lying deep within the forbidden band. The usual impurities present in zinc sulfide are In, Ga, Cl, Br, I, and several other elements. Single crystals of zinc sulfide exhibiting p-type conduction have not yet been prepared. The difficulties encountered in the preparation of p-type crystals are the greater depth of acceptor levels, compared with donors, and the mechanism of charge compensation in which impurities giving rise to acceptor levels require the presence of defects which act as donors [52-56]. Therefore, ZnS is usually regarded as a strongly compensated n-type semiconductor [57].

According to Bube [58], the depth of Cu and Ag acceptor levels in zinc sulfide is 0.9 and 0.6 eV, respectively. These values were obtained as follows: measurements of the impurity excitation edge were used to find the transition energies, which were about 2.8 eV for the Cu levels and about 3.1 eV for the Ag levels. These energies were then subtracted from the optical value of the forbidden band width. According to Bube, all the results obtained in this way should be regarded as approximate because of the Franck–Condon principle. In our opinion, the errors in this approach to acceptor levels in ZnS may be very considerable. An excited electron, transferred from a luminescence center to the conduction band, reaches equilibrium with the lattice and then recombines with an ionized center. Therefore, the transition energy should be subtracted from the thermal value of the forbidden band width, which is 3.2 ± 0.2 eV [2]. It is then found that the depth of the acceptor levels should be about 0.4 eV for copper and about 0.1 eV for silver. Vinokurov and Fok investigated the kinetics of the luminescence emitted by copper-doped ZnS [59] and found that the depth of the copper acceptor level was 0.33 eV. Thus, the ionization energy of the acceptors was approximately equal to the ionization energy of the donors in the ZnS, which were known to lie within the 0.25-0.5 eV range [60, 61]. These results indicate that the depth of the acceptor levels cannot be the basic difficulty in the preparation of p-type ZnS.

Since the bulk conductivity of a semiconductor is governed by carrier transitions from impurity levels and from allowed bands, it follows that the activation energy of impurities and the thermal width of the forbidden band can be found from the temperature dependence of the conductivity. This method of determination of the depth of impurity levels and of the thermal width of the forbidden band was used in investigations of the electrical properties of ZnS [2, 62]. This approach gives the thermal width of the forbidden band of zinc sulfide as equal to 3.2 ± 0.2 eV [2]. The optical width of the forbidden band is 3.7 eV [3, 63, 68]. The conductivity of a semiconductor is governed also by the mobility μ of the free carriers. If the mobility is associated with one type of carrier (either electrons or holes), the relationship between μ and σ can be represented by

$$\sigma = \mu n e,$$

where n is the number of free carriers and e is the electronic charge.

A mobility μ of about 120 cm$^2 \cdot$ V$^{-1} \cdot$ sec^{-1} is reported in [64]. Later investigations [65]   indicate that the electron mobility in ZnS is 80–200 cm$^2 \cdot$ V$^{-1} \cdot$ sec^{-1}. It is worth stressing the good agreement between these values of the electron mobility.

In some cases, the conductivity of zinc sulfide is governed not only by the presence of impurities and by the carrier mobility, but also by the voltage applied to a crystal. The results of Lempicki, Frankl, and Brophy [66] demonstrate that the electrical properties of ZnS crystals grown from the vapor phase are strongly anisotropic: the ratio of the dark conductivities at right–angles to and along the optic axis may reach 10^6 and the corresponding ratio of the photoconductivities may be 10^4. A strong anisotropy of the opposite sign is reported in [67] for crystals grown from the vapor phase. On the other hand, Bochkov, Georgobiani, and Chilaya [62] found that single crystals of ZnS grown by the Stockbarger method are weakly anisotropic. In this case, the ratio of the dark conductivities does not exceed 2–3 and the ratio of the photoconductivities is only 1.5. Bochkov et al. [62] suggest that the strong anisotropy of the electrical properties of the crystals grown from the vapor phase is due to the presence of oriented dislocations.

So far, we have considered the electrical properties of ZnS governed by its bulk characteristics. It must be stressed that the surface conductivity of zinc sulfide may exceed the bulk conductivity by many orders of magnitude. Investigations of the surface conductivity of ZnS and other semiconductors are reported in [69–70]. It has been found that the adsorption of oxygen by n-type semiconductors reduces their surface conductivity. This is explained in [71] by means of a surface state scheme. When an electronegative substance, such as oxygen, is adsorbed chemically in the form of negative ions on the surface of an n-type semiconductor, a net negative charge appears on the surface and this charge is compensated partly or completely by a positive space charge at the surface barrier. Each adsorbed ion effectively removes one free electron from the conduction band. Adsorption thus reduces the surface conductivity. An opposite effect has been reported for p-type semiconductors such as Cu$_2$0. Illumination of a sample affects the adsorption of oxygen on ZnS [69–70], ZnO [72, 73], and CdSe [74].

Contact phenomena are of considerable importance in studies of the electrical properties of semiconductors. Very few systematic investigations have been made of contacts with zinc sulfide. Frankl [75] described some contacts made from a conducting silver dye. The current–voltage characteristics of such contacts are nonlinear. A study of contact barriers on zinc sulfide was made by Alfrey and Taylor [76]. Alfrey and Cooke [77, 78] investigated evaporated and pressure contacts. These contacts were subjected to "electrical forming," consisting of the passage of a strong current. It was found that such forming converted to a nonlinear current — voltage characteristic into a linear or nearly linear dependence.

Such behavior is exhibited by Al, Ga, and In contacts, which give rise to donor levels in zinc sulfide. It has been suggested that heating diffuses these metals into ZnS and produces a surface region with a high donor concentration. This reduces the surface barrier and increases its "transparency" to electrons.

The electrical properties of crystals depend strongly on the chemical binding. For example, the electrical resistivity of ionic semiconductors is usually high and that of covalent materials is normally much lower. However, the distinction between ionic and covalent binding can be applied only to those compounds in which only one type of bond is encountered. Most inorganic substances, including ZnS, have mixed covalent and ionic binding. The type of binding in ZnS was deduced by Browne [79] by calculating the covalent binding energy from the efficiency of luminescence. Browne found that ZnS was mainly ionic because the ionic type of binding amounted to 64%. Estimates of the degree of ionicity based on the electron affinity gave only 30% [80, 81]. The same value was obtained from a comparison of longitudinal and transverse optical phonons [82]. The presence of the ionic component of the binding affects the results of measurements of the forbidden band width. In the thermal method, electrons and holes are in thermodynamic equilibrium with the lattice and move along the lattice in the form of polarons. In the optical method, this equilibrium is disturbed. An optical transition is so rapid that the lattice cannot be polarized in the time available. The polarization (formation of polarons) requires an energy which is represented by the difference between the forbidden band widths obtained by these two methods. In the case of ZnS, this difference amounts to 0.4–0.7 eV [2].

In this section, we have considered mainly the electrical properties of ZnS in darkness. Illumination gives rise to several new phenomena resulting from the interaction of light with matter. Some of these phenomena will be considered in the next section.

§ 2. Photoconductivity and Luminescence

The absorption of photons by bound and free carriers plays an important role in the interaction of light with matter. The absorption by bound carriers, resulting in electron transitions to the conduction band, is manifested primarily by a change in the resistivity. This phenomenon is known as the photoconductivity.

The first measurements of the photoconductivity of pure ZnS crystals were carried out by Piper [83], according to whom the photoconductivity spectrum of pure crystals has one maximum at a photon energy of 3.68 eV. The presence of impurity or defect levels in the forbidden band complicates the spectrum. Electrons may then be excited from the impurity levels to the conduction band, and this gives rise to additional maxima representing the optical activation energies of these levels. Silver, copper, and manganese are the elements used most frequently in the doping of zinc sulfide. All these impurities form luminescence centers in ZnS.

The photoconductivity spectra can be used to determine the nature of the electron transitions associated with the absorption of light. The excitation of Cu and Ag centers results in the generation of free electrons and, therefore, in electron transitions from the impurity levels to the conduction band. The absorption of light of wavelengths corresponding to the luminescence of Mn does not result in the generation of free carriers. This means that the Mn centers become excited. The long- and short-wavelength edges of the excitation spectra of the photoconductivity of ZnS doped with Cu and Ag are in good agreement with the edges of the absorption spectrum and of the luminescence excitation spectrum. Copper-doped ZnS exhibits the following luminescence bands [84–86]: a blue band at 440–450 nm, a green band at 515–530 nm, a yellow band at 570–590 nm, and a red band at 670–720 nm. Silver-doped zinc sulfide has the following luminescence bands [86]: an ultraviolet band at 380–395 nm, a blue band at 435–450 nm, a blue-green band at 480–495 nm, and an orange band at 620–640 nm.

CHAPTER III

FABRICATION AND INVESTIGATION OF p−n JUNCTIONS

§1. Fabrication of p−n Junctions

A. Initial Materials. Single crystals of zinc sulfide grown from the vapor phase were used as the initial material in the fabrication of p−n junctions. Such crystals were doped with chlorine during their growth. The n-type nature of the conduction in these crystals was indicated by the magnitude and sign of the thermoelectric power, determined by the probe method. Chlorine gave rise to donor levels whose depth was about 0.25 eV [60]. The influence of the bulk resistance of ZnS, which was in series with the p—n junction resistance, on the current—voltage characteristics, was minimized by using samples of the lowest possible resistivity. We selected ZnS:Cl samples whose dark resistivity was less than 10^8 $\Omega \cdot$ cm.

A p-type region in such crystals can be formed by overcompensating donor impurities with acceptors. Copper, silver, gold, platinum, etc. can be used as the compensating impurities. We used copper and silver to form p—n junctions in ZnS. These elements were selected because they produced relatively shallow acceptor levels (0.3 eV in the case of Cu [59] and even less in the case of Ag) and because they formed luminescence centers.

B. Method Used to Fabricate p—n Junctions. The diffusion method seemed to be the most suitable in our case. The method of doping during growth from the melt was unsuitable because we started with doped ZnS:Cl single crystals. The alloying method could not be used because of the high melting point of ZnS ($\gtrsim 1800°$C at an excess pressure of some inert gas). Moreover, the melting had to be local and this would have added further complications. For these reasons, we fabricated p—n junctions by diffusion from solid (copper) and liquid (silver) phases.

We selected ZnS:Cl samples which did not have undesirable barriers. Such barriers would have affected the bulk photo-emf and electroluminescence. The initial crystals were selected as follows. Slabs cut from ZnS single crystals were oriented perpendicularly to the growth axis. These slabs were ground and polished. Their thickness was within the range 0.8–1.5 mm and their width did not exceed 3–3.5 mm. Slabs prepared in this way were etched in a mixture of sulfuric acid and potassium dichromate at 90°C for 15–30 min. The composition of the etchant was as follows: 39.2 g H_2SO_4 and 14.7 g $K_2Cr_2O_7$ per 100 ml distilled water. The slabs were then washed in ethyl alcohol and in doubly distilled water at room temperature.

Indium contacts were attached using the method described by Alfrey and Cooke]77, 78]. For reasons to be given later, the contacts were deposited on both sides. The evaporation and the annealing of the indium were carried out in an outgassed vacuum evaporator of the BA-500 type manufactured by Balzers. A slab of ZnS was placed in a special holder (Fig. 8) which — together with a boat containing indium — was kept in the vacuum chamber during the outgassing. This holder could be used to heat a crystal to 700°C. The temperature was measured with a Chromel — Alumel thermocouple. The vacuum chamber containing the ZnS slab was evacuated down to at least 10^{-6} mm Hg. The chamber was cleaned by a glow discharge in an argon atmosphere and the chamber was pumped down to the lowest vacuum attainable. The surface of the ZnS was cleaned by bombardment with argon ions for 40 min. The indium was evaporated from a molybdenum boat in vacuum, which was at least 10^{-6} mm Hg. The distance between the boat and the slab was 75–80 mm. Immediately after its evaporation, the indium was annealed for 5–7 min at 600–650°C. The crystal was heated by a molybdenum strip (Fig. 8), through which an electric current was passed. We then selected those samples which did not exhibit an appreciable electroluminescence at voltages up to 1 kV and did not generate significant photo-emf when illuminated with a PRK-5 lamp.

The indium contacts were removed by grinding and the surface was subjected to a second treatment (etching, washing in ethyl alcohol and doubly distilled water, cleaning by ion bombardment). The copper was then deposited by evaporation in high vacuum (better than 10^{-6} mm Hg). Molybdenum or tantalum boats were used to vaporize the copper. The distance from the boat to the slab was 180–200 mm. The rate of deposition of the copper was 8–12 Å/min. A film whose thickness was at least 400 Å was evaporated in this way. The rate of evaporation was then increased to 150 Å/min and continued for a further 15 min. The temperature of the sample during evaporation was maintained at 280–320°C. Immediately after the evaporation of the copper, the temperature was raised to 620–660°C and maintained at this level for

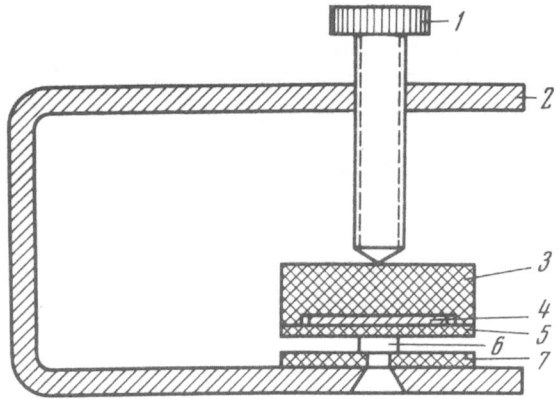

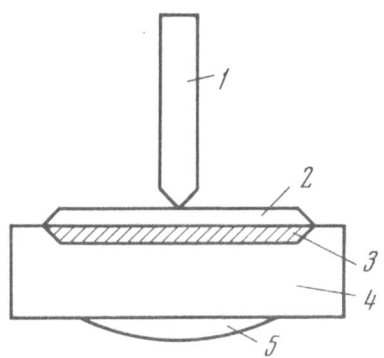

Fig. 8. Holder for mounting and heat treatment: 1) stainless steel screw; 2) stainless steel bracket; 3) ceramic plate; 4) molybdenum strip (d = 0.5 mm); 5) quartz plate (0.5-1 mm); 6) ZnS crystal; 7) mica mask (d = 1 mm).

Fig. 9. Schematic representation of a p — n junction: 1) copper or silver pressure electrode; 2) Cu or Ag film; 3) p-type ZnS; 4) n-type ZnS; 5) indium electrode.

several minutes. This was done to ensure favorable conditions for the diffusion of copper in zinc sulfide. Measurements of the thermoelectric power of the samples doped in this way indicated that they had p-type conduction. In these measurements, the residue of copper as well as the compounds probably formed during the annealing were removed from the surface and the sample was washed in a solution of potassium cyanide. This treatment did not damage the copper-enriched surface region in ZnS [87]. Thus, the p-type region had the composition ZnS:Cu:Cl and the n-type region ZnS:Cl. After its last heat treatment, the sample was immersed for 20-30 sec in an etchant to clean the free surfaces. This naturally etched the copper film which remained after the annealing (this film was used as the contact with the p-type ZnS). Therefore, the thickness of the evaporated copper film was selected in such a way that at least some copper remained on the surface after completion of the full preparation cycle. The thickness of the indium contact with n-type ZnS was selected on the same basis. Figure 9 shows schematically one of the investigated p — n junctions. The junction area was about 10^{-2} cm^2.

The current — voltage and other characteristics of the p — n junctions prepared in this way should not be determined over a sufficiently wide range of temperatures. The results were reproducible only up to 350°K. This was probably because the values of the surface and the bulk diffusion coefficients of copper were high at elevated temperatures and in strong electric fields. This resulted in irreversible spreading of the p — n junction region. Therefore, we decided to fabricate p — n junctions using an acceptor impurity which diffused much more slowly than copper. The most suitable element was silver. This metal was used to fabricate p — n junctions in ZnS in approximately the same way as that described for copper. Since silver diffused much more slowly than copper, the annealing temperature after the evaporation of the silver was raised to 1000°C and the annealing duration was increased to 1 h. The annealing treatment was carried out in an atmosphere of argon from which oxygen and water vapor had been removed (argon was necessary because zinc sulfide sublimated at 1000°C in vacuum). An indium contact with the n-type region was made after the fabrication of a p—n junction. This order was adopted because such a contact could not withstand the temperatures used in the diffusion of silver. The junctions prepared by doping with silver operated over a wide range of temperatures. Fully reproducible results were obtained right up to 800°K.

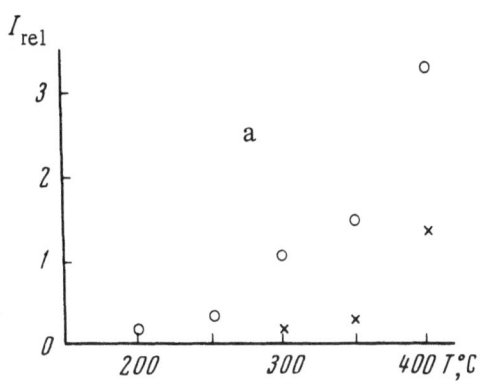

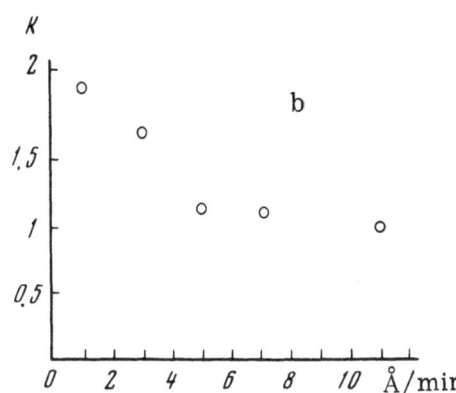

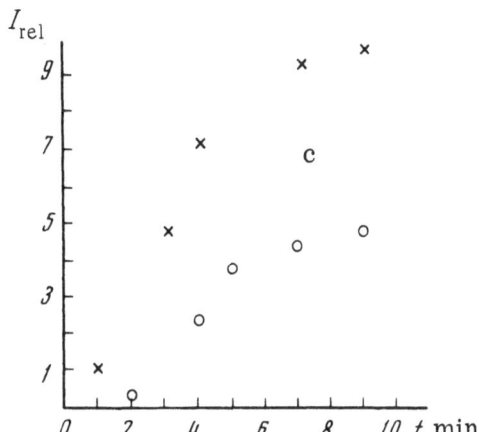

Fig. 10. a) Dependence of the relative intensity of the brightness wave peaks, corresponding to the forward (circles) and reverse (crosses) voltages, on the temperature of a crystal during the evaporation of copper. b) Dependence of the ratio of the brightness wave peaks, corresponding to the forward and reverse voltages, on the rate of evaporation of copper on a crystal heated to 300°C. c) Dependence of the relative intensity of the brightness wave peaks, corresponding to the forward (crosses) and reverse (circles) voltages, on the duration of annealing during p−n junction fabrication.

These samples not only had good rectifying characteristics but also exhibited electroluminescence under forward and reverse voltages.* This point will be considered in more detail in a later section.

C. Selection of Optimal Fabrication Conditions. We selected the optimal conditions for the fabrication of p−n junctions in ZnS on the basis of the brightness waves observed in the luminescence under forward and reverse voltages. These brightness waves represented variations of the luminescence intensity during each period of the external alternating voltage. The brightness waves were observed at a voltage of 100 V of 100 Hz frequency. Their nature was basically the same for other voltages and frequencies (at least up to 1000 V and 8 kHz).

The temperature of a crystal during the deposition of the copper was selected to ensure the highest values of the amplitudes in the brightness oscillograms corresponding to the forward and reverse voltages. The relative intensity I_{rel} of the peaks of the brightness waves emitted by a p−n junction is plotted in Fig. 10a as a function of the temperature of a crystal during the evaporation of the copper. The evaporated film was annealed for 5 min at 650°C. The evaporation process was carried out in two stages: (1) 50 min at a rate of 8 Å/min on a heated crystal; (2) 10 min at a rate of 150 Å/min on the same crystal during the cooling stage. The luminescence under forward voltages decreased considerably when the rate of evaporation during the first stage was increased.

---

* Since the luminescence characteristics were of special interest, we selected those p−n junctions which exhibited the strongest luminescence under forward and reverse voltages.

It is evident from Fig. 10a that the higher the temperature of a crystal during the evaporation of copper, the stronger is the luminescence of a p — n junction. However, the films evaporated at 400°C have uneven thicknesses and this affects considerably the reproducibility of the results. The re-evaporation of copper from the surface of a crystal is also likely to be considerable at high temperatures. To some extent, the results may be affected by the distribution of the temperature across the surface of a crystal. We maintained the temperature of a crystal at 280-320°C during the evaporation stage. This ensured that our results were reproducible.

Figure 10b shows the dependence of the ratio of the brightness wave peaks, corresponding to the forward and reverse voltages across a p—n junction, on the rate of deposition of copper on the surface of a crystal kept at a temperature of about 300°C. The copper was evaporated in two stages: 1) slow evaporation at a rate of 8 Å/min on a heated crystal until a thickness of at least 400 Å was achieved; 2) the heater was switched off when this thickness was reached and the rate of evaporation was increased to 150 Å/min and continued for a further 10 min.

When the rate of evaporation was reduced, the intensity of the luminescence under forward voltages increased relative to the intensity obtained under reverse voltages. When the copper was deposited very slowly, the luminescence under a forward bias predominated. Since our intention was to investigate the luminescence generated by forward and reverse voltages, we selected an evaporation rate which would ensure that the intensities of the luminescence under forward and reverse voltages were comparable. The rate of evaporation which gave this result was within the range 7-12 Å/min. We were unable to give a satisfactory explanation of the observed dependence of the relative intensities of the brightness wave peaks on the rate of evaporation of the copper because we did not have information on the

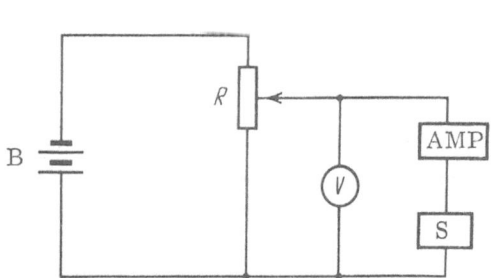

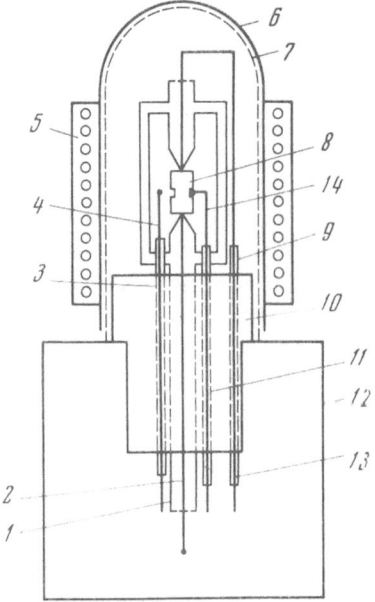

Fig. 11. Circuit used in measurements of the current — voltage characteristics: B is a battery; R is a rheostat; V is a voltmeter of the VLU-2 type; AMP is a dc amplifier of the U1-2 type; S is the sample under investigation.

Fig. 12. Light-tight thermostat: 1) quartz tube; 2) platinum electrode; 3) quartz tube; 4) thermocouple; 5) heater; 6) quartz chamber; 7) metal screen; 8) sample being investigated; 9) quartz tube; 10) platinum electrode; 11) quartz tube; 12) screen; 13) platinum electrode.

interaction between the copper atoms and the surface or bulk of the ZnS. When the duration of the slow evaporation stage was reduced, the luminescence of a p—n junction under a forward bias decreased considerably, or even disappeared.

The electroluminescent properties of p—n junctions are affected also by the duration of annealing. Figure 10c shows the dependence of the relative intensity of the brightness peaks, corresponding to the forward and reverse directions, on the duration of annealing. It is evident that the amplitudes of the brightness waves obtained under forward and reverse voltages reached saturation for crystals annealed for about 6-7 min. When the annealing duration was extended to at least 20 min, the intensity of the forward luminescence decreased considerably. Therefore, the annealing duration in our experiments did not exceed 10 min and was usually 4-7 min.

§2. Methods Used in Investigations of p—n Junctions in ZnS

A. Recording of Current — Voltage Characteristics. The circuit used to record the current — voltage characteristics is given in Fig. 11. The current flowing through a junction was measured using a dc amplifier of the U1-2 type. The sensitivity of this amplifier was about 10^{-12} A. The sample being investigated was placed in a light-tight thermostat (Fig. 12). Before the measurements were made, the sample was de-excited by brief illumination with red light. The temperature was measured with a copper — constantan thermocouple. The high-temperature measurements on p—n junctions fabricated by the diffusion of silver were carried out in an atmosphere of argon from which oxygen and water vapor had been removed.

B. Excitation and Recording of Photo-Emf. The photo-emf of p—n junctions in ZnS was excited as shown in Fig. 13. The emf was recorded using the same dc amplifier (set to measure the emf) as in the measurements of the current—voltage characteristics. The maximum input resistance of the amplifier was 10^{11} Ω. The measurements were carried out by direct and compensation methods using the U1-2 as a null indicator. Both methods gave the same results. Excitation was provided by a mercury lamp of the PRK-5 type, whose radiation was passed through a ZMR-3 monochromator.

C. Excitation and Recording of Electroluminescence. Sinusoidal and static voltages were used to excite the electroluminescence. Audiofrequency oscillators of the

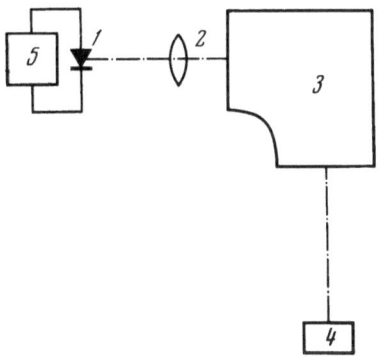

Fig. 13. Unit used in the excitation of photo-emf: 1) sample; 2) quartz lens; 3) ZMR-3 monochromator; 4) PRK-5 mercury lamp; 5) U1-2 dc amplifier.

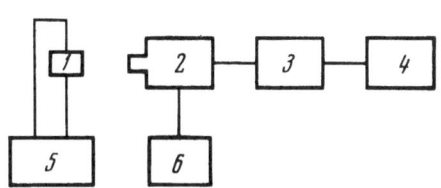

Fig. 14. Apparatus used to excite and record the electroluminescence emitted by p—n junctions: 1) sample; 2) photomultiplier FÉU-36; 3) amplifier; 4) recorder; 5) excitation voltage source; 6) photomultiplier power supply.

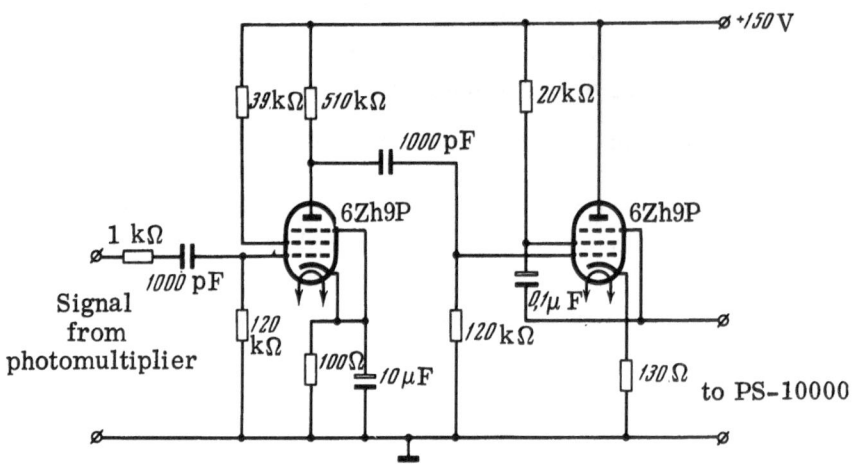

Fig. 15. Pulse amplifier circuit.

ZG–12 and GZ–33 type were used as the sinusoidal voltage sources. Batteries were employed to provide a static voltage. As mentioned in the preceding section, the optimal conditions for the fabrication of electroluminescent p—n junctions in zinc sulfide were selected on the basis of the brightness waves. These waves were observed in a specially constructed unit, shown schematically in Fig. 14. This unit included a photomultiplier (FÉU-36) supplied with a static voltage of 1900 V from a high-voltage source of the IVN-1 type. The signal taken from the photomultiplier was boosted by the preamplifier stage of a wide-band amplifier of the USh-2 type. The brightness waves were observed on the screen of a double-beam oscillograph of the D-581-type.

A photometric unit was used in the measurements of the integrated intensity in order to count the incident quanta. In this way, we were able to record even a few quanta of visible radiation per second. The dark current of a special photomultiplier was reduced by placing it in a liquid-nitrogen dewar. A PS-10000 counter was used as the detector. The sensitivity of this counter was 10^{-3} V and its resolution time was 1 μsec. The counting operation was stopped automatically. At high luminescence intensities, we used an M-95 microammeter, capable of measuring collisions down to 2×10^{-9} A. The pulse amplifier (Fig. 15) was based on a 6Zh9P tube. The cathode follower, also based on a 6Zh9P tube, was used to isolate the detector from the amplifier. A stabilized rectifier of the VS-16 type was used to supply the photomultiplier. The electroluminescence spectra of the p—n junctions were recorded using a Hilger quartz spectrograph and a very sensitive type 13 film (as used in aerial photography), or a film of the RT-3 type. The same spectrograph was employed to record the photoluminescence spectra of the p—n junctions excited with the λ = 313 nm mercury line.

CHAPTER IV

OPTICAL AND ELECTRICAL PROPERTIES OF p—n JUNCTIONS IN ZnS

§1. Current — Voltage Characteristics and Photo-emf

The current—voltage characteristics of ZnS diodes were recorded in the range 290–350°K for the copper-doped diodes and in the range 290–800°K for the silver-doped diodes. The experimental results were analyzed on the basis of the Shockley theory [5] and the Sah—Noyce—Shockley theory [8]. Static current-voltage characteristics, recorded at 350°K, are given in

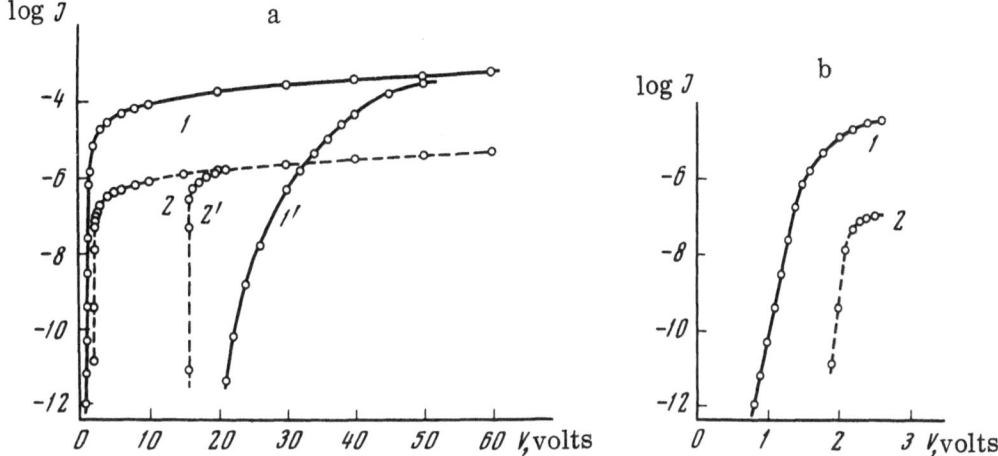

Fig. 16. Current—voltage characteristics of p—n junctions in ZnS, fabricated by the diffusion of copper (1, 1') or silver (2, 2'). a: 1), 2) Forward branches; 1'), 2') reverse branches. b: Initial parts of curves 1 and 2.

Fig. 16. The abscissa represents the voltage and the ordinate the logarithm of the current expressed in amperes. It is evident from this figure that the dependence of log J on V is linear up to 1.5 V for the copper-doped diode and up to 2.1 V for the silver-doped diode. This means that the current flowing through a p—n junction increases exponentially with the forward voltage. The linear regions in Fig. 16 can be approximated satisfactorily by the Sah—Noyce—Shockley formula [8], which is of the following form eV ≫ kT:

$$J = J_s \exp \frac{eV}{AkT}. \qquad (4')$$

Equation (4') is derived taking account of carrier recombination in the p—n junction. If all the minority carriers recombine in the junction, the factor A has its maximum value of 2; if there is no recombination, we find that A = 1 (the Shockley formula for eV ≫ kT). The current — voltage characteristics of p—n junctions, shown in Fig. 16, yield the following values of A: 1.08 ± 0.08 for the silver-doped diode and 1.95 ± 0.07 for the copper-doped diode. These values of A indicate strong recombination in the p—n junction region in the copper-doped diode and practically no recombination in the silver-doped diode. Nevertheless, the silver-doped diode emits electroluminescence at low forward voltages. It is likely that, in this case, carriers recombine outside the junction (in the bulk ZnS) and the current through the junction is due to carrier diffusion.

At voltages above 3 V, the rise of the current slows down very appreciably, which is probably due to the influence of the bulk resistance. It follows that the p—n junction barrier

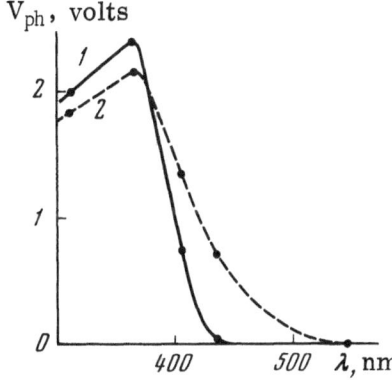

Fig. 17. Dependence of the photo-emf on the wavelength of the exciting radiation: 1) silver-doped diode; 2) copper-doped diode.

is 1.5 < ψ ≤ 3 eV.   The maximum value of the photo-emf (Fig. 17) excited with a powerful mercury lamp is 2.4 and 2.15 V for the silver- and copper-doped diodes, respectively. Since the maximum value of the photo-emf is equal to or less than the barrier height, it follows that this height is within the range 2.4 ≤ ψ_{Ag} < 3 eV, or 2.15 ≤ ψ_{Cu} < 3 eV. The photo emf spectra shown in Fig. 17 have maxima at an excitation wavelength of about 365 nm. The photo-emf decreases at the short-wavelength end, as in the case of p−n junctions in ZnSe. It is probable that this fall is due to weak penetration of the short-wavelength radiation into zinc sulfide.

The current−voltage characteristics, plotted on a semilogarithmic scale (Fig. 18), have linear regions at all the temperatures used in the present investigation. These regions can be described by Eq. (4') with A = 1-2. At high temperatures, the silver-doped diodes exhibit a kink in the low-voltage part of the forward current−voltage characteristic and this gives rise to two regions with different values of A. At low voltages, we have 2 > A > 1, but when the voltage is increased, the slope changes and A decreases to 1.08 ± 0.08, i.e., it effectively becomes unity. According to the Sah−Noyce−Shockley theory, this value of the coefficient A indicates that carrier diffusion across the p−n junction predominates over carrier recombination in the junction. The linear region corresponding to A ≈ 1 is described easily by the Shockley equation [see Eq. (1) and [5]]. The kink in the initial part of the forward branch of the current−voltage characteristics of the silver-doped diodes may be attributed to the effect of the p−n junction field on carrier diffusion. At voltages lower than the barrier height, the junction field is sufficiently strong to retain a considerable fraction of carriers in the junction region and thus enhance their recombination so that the coefficient A increases. The pre-exponential factor $\mathcal{J}_s$ in Eq. (4') is of the form

$$\mathcal{J}_s \propto \exp\left(-\frac{E_T}{AkT}\right).$$

Here, E_T is the thermal width of the forbidden band of the semiconductor. The temperature dependence of this factor can be used to determine the thermal width of the forbidden band. The value of $\mathcal{J}_s$ is found by extrapolation of the linear regions in the semilogarithmic current−voltage characteristics, until they intersect the current axis. Figure 19a shows the temperature dependences of log $\mathcal{J}_s$ obtained in this way for the copper- and silver-doped diodes

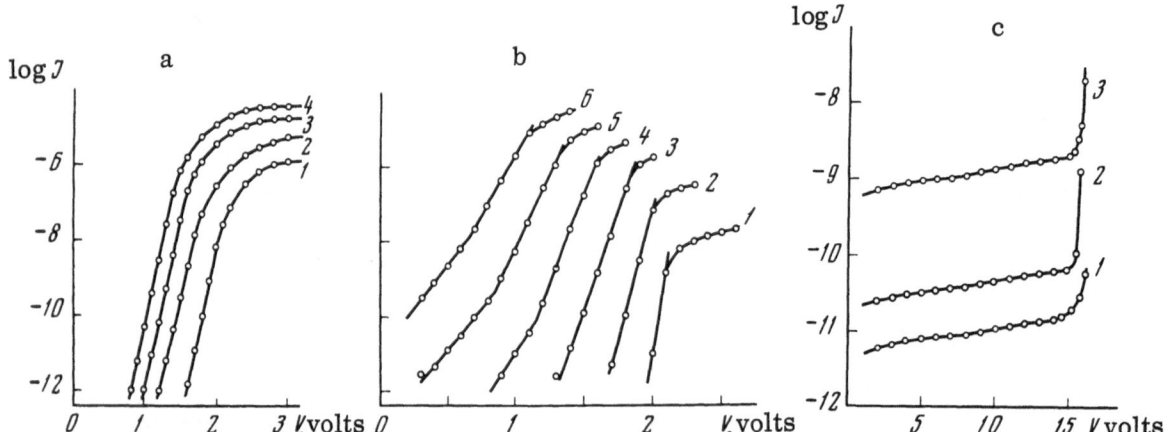

Fig. 18. Forward (a, b) and reverse (c) branches of the current−voltage characteristics of p−n junctions at various temperatures. a) Copper-doped diode: 1) 290, 2) 310, 3) 330, 4) 350°K; b) silver-doped diode: 1) 300, 2) 400, 3) 500, 4) 600, 5) 700, 6) 800°K; c) silver-doped diode: 1) 700, 2) 750, 3) 800°K.

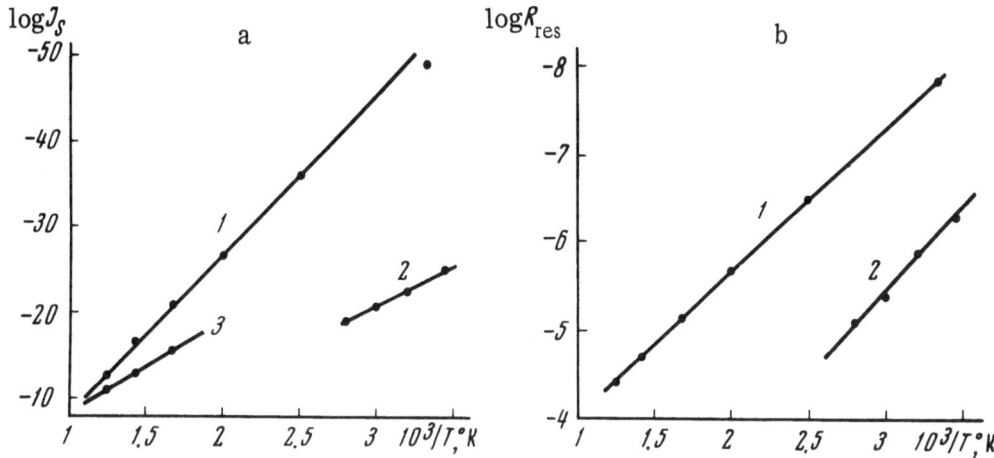

Fig. 19. a) Temperature dependence of $\mathcal{J}_s$ for silver-doped (1, 3) and copper-doped (2) diodes. b) Temperature dependence of the residual resistance of silver-doped (1) and copper-doped (2) diodes.

(line 1 corresponds to the steeper linear region and A = 1, whereas line 3 corresponds to the less steep region — see Fig. 18b). The slopes of the dependences 1 and 2 in Fig. 19a were used to find E_T for ZnS: E_T = 3.42 ± 0.09 eV. This value corresponded to T = 0°K and was in good agreement with the results reported in [2] .

The forward current flowing through a ZnS diode is limited by the bulk resistance of the original crystal. An investigation of the temperature dependence of the current — voltage characteristics and a determination of the residual resistance give the ionization energy ΔE of the impurities responsible for the bulk conductivity of ZnS, in accordance with the following formula:

$$R_{\text{res}} = R_0 \exp\left(-\frac{\Delta E}{kT}\right),$$

which is valid for $n \ll N_a$, where n is the number of free electrons in the original (n-type) part of the crystal and N_a is the density of the acceptor levels in that part of the crystal.

Figure 19b shows the temperature dependence of the residual resistance of the ZnS diodes. It is evident from this figure that the temperature dependence of R_{res} can be represented by a straight line if plotted using the coordinates log R_{res} and 1/T°K. The ionization energy ΔE found from the slopes of the straight lines in Fig. 19b is 0.35 ± 0.1 eV and corresponds to the level of Cl (this value is in agreement, within the limits of experimental error, with 0.25 eV [60]). The reverse branches of the current — voltage characteristics (1' and 2' in Fig. 16a) have regions of rapid rise of the current. The voltage at which this rise occurs differs from one diode to another and lies within the range 12-30 V. The rise is more rapid for the silver-doped diodes. At higher reverse voltages, the rise of the current slows down (this is probably associated with the breakdown of the p—n junction). The high-voltage peaks of the current — voltage characteristics are naturally strongly affected by the bulk resistance of a crystal. The reverse branches of the current — voltage characteristics of the silver-doped diodes shown in Fig. 18c were obtained at high temperatures (700, 750, and 800°K). It is evident from Fig. 18c that the reverse current does not reach saturation but increases continuously with increasing applied voltage. This behavior of the current may be attributed to the expansion of the space charge region and is not in conflict with the Sah — Noyce — Shockley theory.

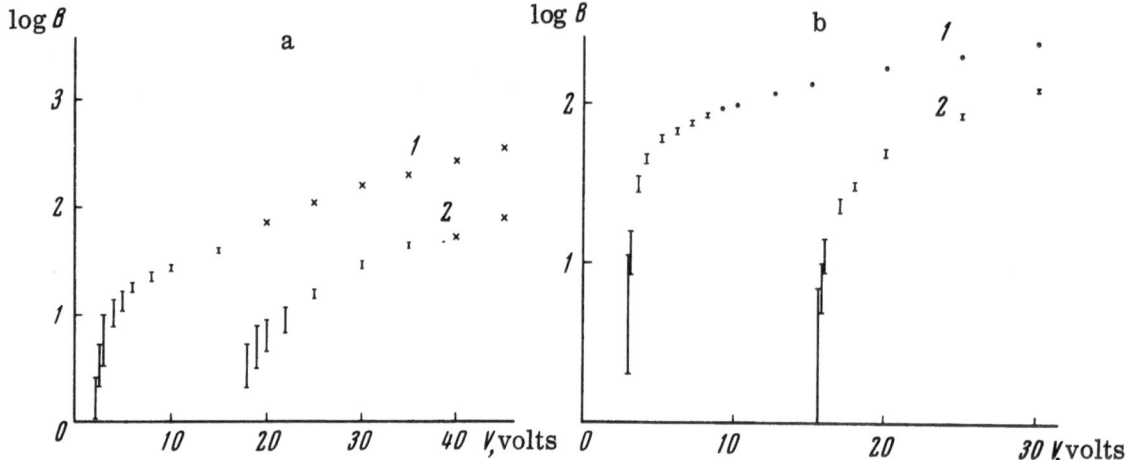

Fig. 20. Dependence of the electroluminescence brightness on the applied voltage: 1) forward voltage; 2) reverse voltage; a) copper-doped diode; b) silver-doped diode.

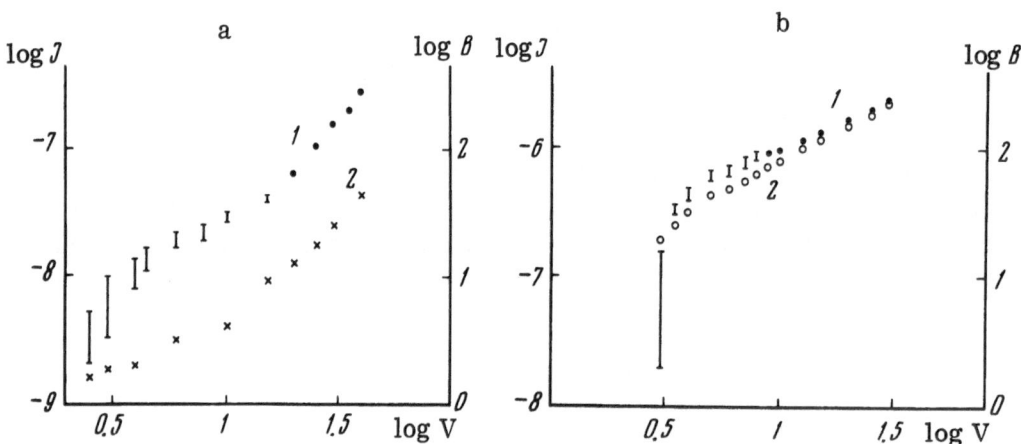

Fig. 21. Dependence of the current flowing through a p — n junction (2) and of the electroluminescence brightness (1) on the applied voltage: a) copper-doped diode; b) silver-doped diode.

§2. Electroluminescence of p — n Junctions in ZnS

Our p — n junctions emitted electroluminescence under forward and reverse voltages. The luminescence was localized in the p — n junction region. Under forward voltages, visible luminescence was recorded at 2.2 V in the case of the copper-doped diodes and at 2.9 V in the case of the silver-doped diodes. The dependence of the luminescence brightness B on the applied voltage is presented in Fig. 20. The ordinate is the logarithm of the brightness in relative units. The appearance of luminescence at these very low voltages may be due to the injection of minority carriers through a p — n junction. The injection mechanism is supported also by the linear dependence of the brightness on the current through the junction. Figure 21 shows the dependences of the forward current and of the luminescence brightness on the voltage applied to a copper-doped and silver-doped diode. The dependences are plotted on a double logarithmic scale.

Appreciable luminescence under a reverse voltage (Fig. 20) is observed when this voltage reaches values corresponding to the strong rise of the current through the diode (this rise is due to the breakdown of the p — n junction).

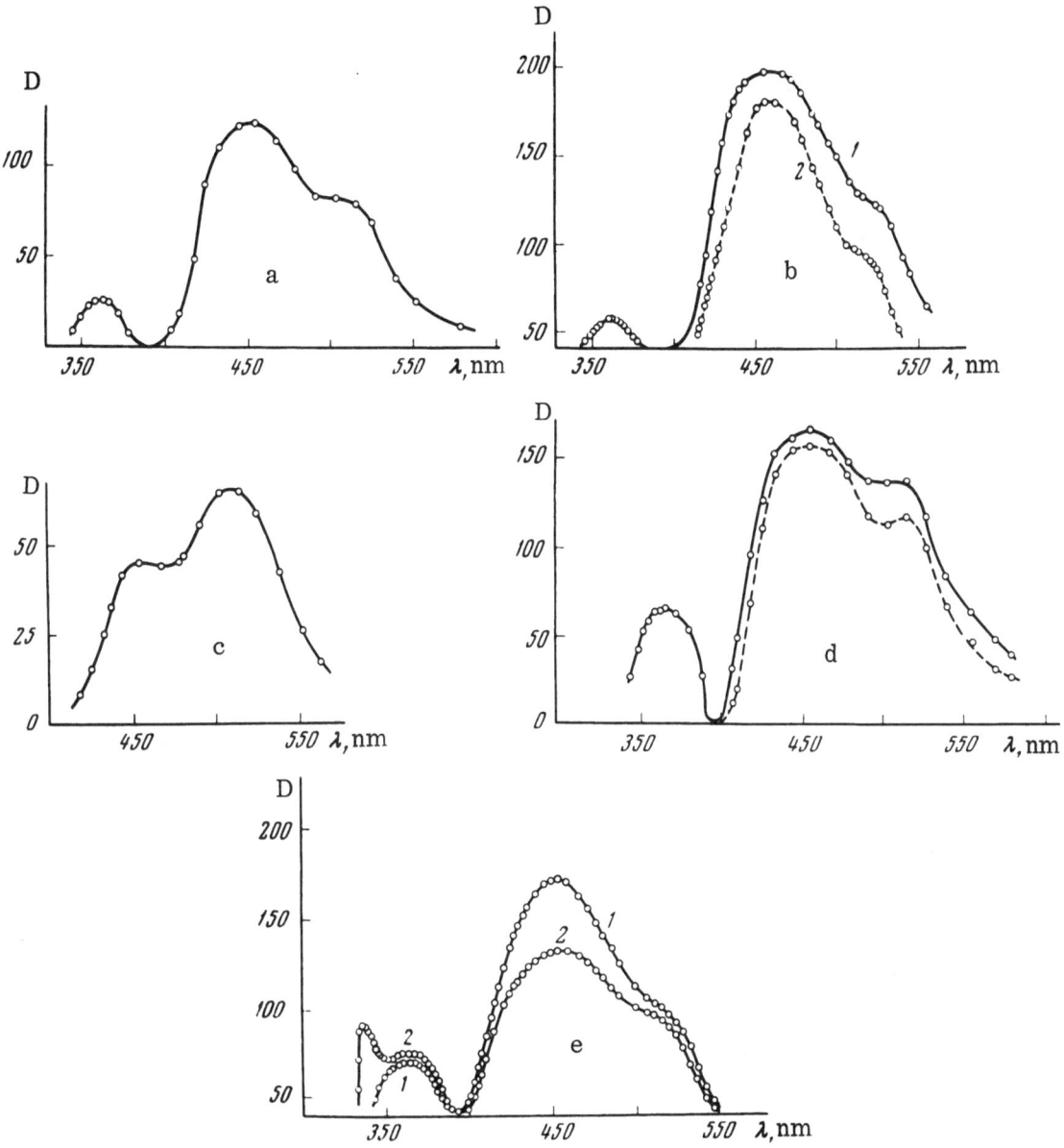

Fig. 22. Microphotograms of: a) electroluminescence spectrum of a copper-doped diode subjected to 5 V in the forward direction; b) luminescence spectra of a silver-doped diode subjected to electrical (5 V − curve 1) and optical (curve 2) excitation; c) photoluminescence spectrum of a copper-doped diode; d) electroluminescence spectra of a copper-doped diode subjected to forward (continuous curve) and reverse (dashed curve) voltages of 25 V; e) electroluminescence spectra of a silver-doped diode subjected to forward (1) and reverse (2) voltages of 25 V.

The electroluminescence and photoluminescence spectra were determined for the same samples. The electroluminescence spectra were recorded at voltages of 5 and 25 V applied in the forward and reverse directions. At 5 V, the electroluminescence was observed only for the forward direction. At 25 V, the junctions emitted electroluminescence for both directions of the voltage (Fig. 22).

It is evident from Fig. 22 that the luminescence spectrum obtained under a forward voltage of 5 V consists of three bands: a "green" band with a maximum at about 510 nm, a "blue" band with a maximum at about 450 nm, and an ultraviolet band with a maximum at about 360 nm. Only the long-wavelength bands were found in the photoluminescence spectra. These bands probably represented the usual recombination of electron—hole pairs at the luminescence centers.

When 25 V was applied in the forward or reverse direction, the luminescence bands in the spectra of the copper-doped diodes were the same as the bands observed under a voltage of 5 V but the ultraviolet band was observed only for the forward direction. The silver—doped diodes behaved differently. When they were subjected to a reverse bias, they emitted two ultraviolet bands with maxima at 360 and 335 nm. The short-wavelength edge of the 335 nm band was steep probably because of the fundamental absorption by the crystal lattice. Since the short-wavelength ($\lambda_{max} \approx 335$ nm, $h\nu_{max} = 3.7$ eV) corresponded to the optical width of the forbidden band, we attributed this maximum to the interband recombination of electrons and holes which were not in the polaron state. A major part of the ultraviolet peak at $\lambda_{max} = 360$ nm ($h\nu_{max} = 3.44$ eV), as well as its maximum, was found to be located in the energy interval between the optical and thermal forbidden band widths. For this reason, most of this peak could also be attributed to the interband recombination of electrons and holes but — in contrast to the short-wavelength peak — either one or both recombination partners were in the polaron state. Such ultraviolet luminescence bands have been reported in [88] in the case of electrical breakdown of uniform samples of pure zinc sulfide.

§3. Comparison of Some Characteristics of p—n Junctions in ZnS with the Characteristics of ZnS: Cu_2S Heterojunctions

In spite of the fact that Cu_2S and Ag_2S were dissolved in a potassium cyanide solution, one could argue that the p-type conduction of the copper- or silver-doped zinc sulfide was due to cuprous sulfide or silver sulfide, which were not removed by this treatment. This objection can be overcome by preparing ZnS:Cu_2S heterojunctions and by comparing some of their characteristics with those of p—n homojunctions [89].

Such heterojunctions were prepared as follows: ZnS was polished and etched in a mixture of sulfuric acid and potassium dichromate, cleaned by ion bombardment, and coated with cuprous sulfide by evaporation in vacuum (better than 10^{-6} mm Hg). Measurements of the thermo-

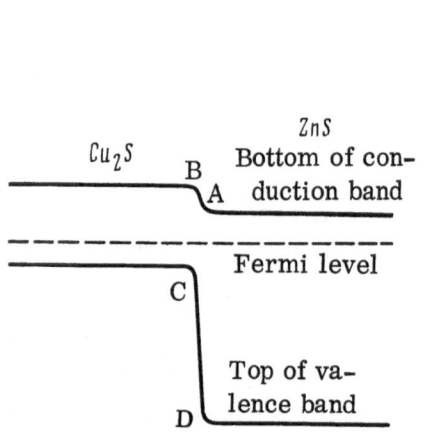

Fig. 23. Energy band scheme of a ZnS:Cu_2S heterojunction in the absence of an external voltage.

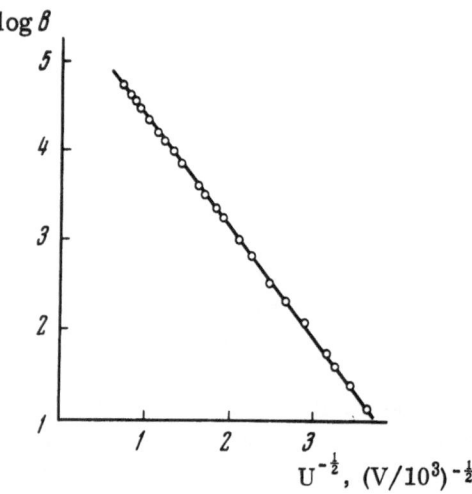

Fig. 24. Dependence of the electroluminescence brightness on the voltage applied to a ZnS:Cu_2S heterojunction.

electric power indicated that the Cu_2S film had p-type conduction and the ZnS substrate n-type conduction. During the deposition of Cu_2S, the ZnS substrate was kept at a low temperature to avoid the undesirable diffusion of copper from the Cu_2S into the ZnS. This system corresponded very closely to the $ZnS:Cu_2S$ heterojunction band scheme shown in Fig. 23.

Figure 24 shows the dependence of the brightness of the electroluminescence, emitted by such a heterojunction, on the reverse voltage. The reverse voltage in the case of a heterojunction is that voltage under which the ZnS substrate is positive and the Cu_2S film negative. The luminescence of such a heterojunction is blue and localized in the junction region. The dependence of the brightness on the applied voltage is linear if plotted in terms of the coordinates $\log B$ and $1/\sqrt{U}$ (Fig. 24). Consequently, this dependence can be represented by a formula which is usually valid for powdered electroluminescent phosphors:

$$B = B_0 \exp\left(-\frac{b}{\sqrt{V}}\right),$$

where B is the brightness, V is the applied voltage; B_0 and b are quantities independent of the voltage.

According to the proposed band scheme, no electroluminescence should be observed when a heterojunction is subjected to a forward voltage. Such electroluminescence would require the injection of holes from Cu_2S into ZnS. However, the valence band of the heterojunction has a barrier CD (Fig. 23) which prevents such injection. When an electric field is applied, the height of this barrier decreases and the electron barrier AB in the conduction band also decreases. Since the barrier AB is lower than the barrier CD, a much lower voltage is required to destroy the former. When the barrier AB in the conduction band is suppressed, the electron current shunts the junctions and prevents the concentration of the electric field until fields capable of destroying the hole barrier are reached. The $ZnS:Cu_2S$ barriers subjected to a forward voltage did not emit electroluminescence up to 500 V although our photometric apparatus was capable of recording even a few quanta per second. The emission of electroluminescence under reverse voltages indicated the existence of luminescence centers at the $ZnS:Cu_2S$ interface and, therefore, luminescence would have been observed had holes been injected into the ZnS under a forward bias.

However, Aven and Cusano [90, 91] found that a $ZnS:Cu_2S$ heterojunction luminesced when subjected to a forward voltage. Aven and Cusano assumed the existence of a triangular barrier (Fig. 25) which prevented the shunting of the junction by the electron current and made it possible to increase the field concentration within the junction until the barrier CD was suppressed. In contrast to our precautions, Aven and Cusano [90, 91] prepared their heterojunctions under conditions favoring the diffusion of copper across the $ZnS:Cu_2S$ interface. Our heterojunctions also began to electroluminesce under a forward voltage if they were subjected to heat treatment before the measurements were made. In our opinion, such heat treatment (annealing) caused the copper to diffuse into the zinc sulfide and produced a p-type region similar to that found in our investigation [92] of p−n junctions prepared from ZnS. We compared the electrical and optical characteristics of the junctions prepared by the evaporation of copper and Cu_2S on ZnS, followed by

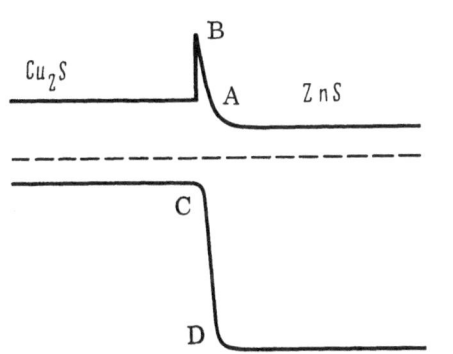

Fig. 25. Energy band scheme of a $ZnS:Cu_2S$ heterojunction.

thermal diffusion. In both cases, the results were identical. Therefore, we concluded that Aven and Cusano [90, 91] observed the electroluminescence of a p–n junction in zinc sulfide.

The absence of electroluminescence from the ZnS:Cu$_2$S heterojunctions under forward voltages indicated that the postulated triangular barrier was either absent and the scheme suggested in Fig. 23 was valid, or that this barrier was sufficiently narrow to permit electron tunneling.

Hence, we concluded that the electrical and optical properties of our p–n junctions were not due to the presence of ZnS:Cu$_2$S heterojunctions.

CONCLUSIONS

The experimental results reported in the present paper show that p-type conduction can be produced in ZnS by the thermal diffusion of copper or silver.

Diodes prepared by our method have good rectifying characteristics: the rectification coefficient reaches 10^7 and the reverse current under a few volts amounts to $\leq 10^{-12}$ A. The heights of the barriers within p–n junctions (deduced from the forward current–voltage characteristics, the value of the photo-emf, and the appearance of injection electroluminescence) are $2.15 \leq \psi < 3$ eV for the copper-doped diodes, and $2.4 \leq \psi < 3$ eV for the silver-doped diodes. The temperature dependences of the pre-exponential factor $\mathcal{I}_s$ in Eq. (4'), (this factor describes the initial part of the forward current–voltage characteristic) can be used to determine the thermal width of the forbidden band of ZnS: 3.42 ± 0.09 eV (this corresponds to T = 0°K).

When p–n junctions in ZnS are illuminated with short-wavelength visible or ultraviolet radiation, they produce considerable photo-emf's, which may reach 2.4 V for the silver-doped diodes. Such a high photo-emf may be used in the detection of short-wavelength radiation.

The application of a forward or reverse voltage to ZnS diodes produces electroluminescence. The luminescence generated by forward voltages appears at 2.2 V in the case of the copper-doped diodes and at 2.9 V in the case of the silver-doped diodes. Under reverse voltages, the electroluminescence appears only at 10-20 V. The luminescence emerges from the junction region. Forward voltages produce "green" (~ 510 nm), "blue" (~ 450 nm), and ultraviolet (~ 360 nm) luminescence bands. Reverse voltages produce only the "green" and "blue" bands. The same bands are observed also in the photoluminescence spectra of these diodes. Diodes prepared by the thermal diffusion of silver exhibit two ultraviolet peaks at about 335 and 360 nm when these diodes are subjected to reverse voltages. These ultraviolet peaks are due to the interband recombination of electron-hole pairs generated during the breakdown of the p–n junctions. Such diodes need low currents and can therefore be used in radio-frequency circuits operating at "zero" powers (at nanowatt and picowatt levels). By way of comparison, it may be mentioned that mass-produced germanium and silicon diodes require not less than 1 μW.

The p–n junctions fabricated by the diffusion of silver in ZnS retain their good rectifying characteristics up to 500°C. This makes them superior to diodes made of germanium silicon, or gallium arsenide. The characteristics of ZnS diodes are hardly affected by storage in air for several months. Since the current through a ZnS diode is limited by its residual resistance, it follows that the resistance of the orignal zinc sulfide crystal must be reduced considerably if we wish to increase the brightness of the electroluminescence emitted by such a diode. If this problem is solved we should be able to use p–n junctions in zinc sulfide as highly efficient low-voltage sources of visible and ultraviolet radiation.

The authors are grateful to E. I. Panasyuk for kindly supplying the zinc sulfide crystals and to V. S. Vavilov, I. K. Vereshchagin, N. A. Penin, and M. V. Fok for discussing the results and for their valuable comments.

LITERATURE CITED

1. M. V. Fok, Opt. Spektrosk., 18:1024 (1965).
2. M. V. Fok, Fiz. Tverd. Tela 5:1489 (1963).
3. M. Cardona and G. Harbeke, Phys. Rev., 137:A1467 (1965).
4. A. G. Fischer, Solid-State Electron., 2:232 (1961).
5. W. Shockley, Bell Syst. Tech. J., 28:435 (1949).
6. K. B. Tolpygo and É. I. Rashba, Tr. Inst. Fiz. Akad. Nauk Ukr. SSR, 7:60 (1956).
7. V. E. Kosenko, Zh. Tekh. Fiz., 27:452 (1957).
8. C. T. Sah, R. N. Noyce and W. Shockley, Proc. IRE, 45: 1228 (1957).
9. N. A Belova, A. N. Kovalev, and N. A. Penin, Fiz. Tverd. Tela, 2:2647 (1960).
10. D. N. Nasledov and B. V. Tsarenkov, Fiz. Tverd. Tela, Sbornik (supplement), 1: 78 (1959).
11. Yu. M. Burdukov, A. N. Imenkov, D. N. Nasledov and B. V. Tsarenkov, Fiz. Tverd. Tela, 3:991 (1961).
12. T. N. Morgan, M. Pilkuhn, and H. Rupprecht, Phys. Rev., 138:A1551 (1965).
13. V. I. Stafeev, Zh. Tekh. Fiz., 28:1631 (1958).
14. R. Newman, Phys. Rev., 91:1313 (1953).
15. J. R. Haynes and W. C. Westphal, Phys. Rev., 101:1676 (1956).
16. K. Lehovec, C. A. Accardo, and E. Jamgochian, Phys. Rev., 89:20 (1953).
17. D. N. Nasledov, A. A. Rogachev, S. M. Ryvkin and B. V. Tsarenkov, Fiz. Tverd. Tela 4:1062 (1962).
18. M. Gerschenzon and R. M. Mikulyak, J. Appl. Phys., 32:1338 (1961).
19. N. Watanabe, Jap. J. Appl. Phys., 5:12 (1966).
20. H. Lozykowski, Czech. J. Phys., B13:164 (1963).
21. R. Newman, Phys. Rev., 105:1715 (1957).
22. A. A. Gippius and V. S. Vavilov, Radiative Recombination in Germanium, Preprint A-100 [in Russian] , Fiz. Inst. Akad. Nauk SSSR, Moscow (1962).
23. N. Watanabe, Jap. J. Appl. Phys., 4:343 (1965).
24. R. H. Bube and E. L. Lind, Phys. Rev., 110:1040 (1958).
25. R. Braunstein, Phys. Rev., 99:1892 (1955).
26. M. A. Melehy and E. A. Jarmoc, Proc. IEEE, 51:1365 (1963).
27. H. K. Henisch, Electroluminescence, Pergamon Press, Oxford (1962).
28. T. N. Morgan, Phys. Rev., 139:A294 (1965).
29. R. Newman, Phys. Rev., 100:700 (1955).
30. M. Kikuchi and K. Tachikawa, J. Phys. Soc. Jap., 14:1830 (1959).
31. R. A. Logan and A. G. Chynoweth, J. Appl. Phys., 33:1649 (1962).
32. R. A. Logan, A. G. Chynoweth and B. G. Cohen, Phys. Rev., 128:2518 (1962).
33. G. F. Kholuyanov, Fiz. Tverd. Tela, 3:3314 (1961).
34. R. L. Batdorf, A. G. Chynoweth, G. C. Dacey, and P. W. Foy, J. Appl. Phys., 31: 1153 (1960).
35. A. G. Chynoweth and H. K. Gummel, J. Phys. Chem. Solids, 16:191 (1960).
36. A. G. Chynoweth and K. G. McKay, Phys. Rev., 106:418 (1957).
37. A. G. Chynoweth and K. G. McKay, Phys. Rev., 108:29 (1957).
38. A. G. Chynoweth and K. G. McKay, Phys. Rev., 102:369 (1956).
39. P. A. Wolff, J. Phys. Chem. Solids, 16:184 (1960).
40. A. E. Michel, M. I. Nathan, and J. C. Marinace, J. Appl. Phys., 35:3543 (1964).
41. A. E. Michel and M. I. Nathan, Bull. Amer. Phys. Soc., 9:269 (1964).

42.   K. Nojima and S. Ibuki, Jap. J. Appl. Phys., 5:253 (1966).

43.   S. G. Ellis, F. Herman, E. E. Loebner, W. J. Merz, C. W. Struck, and J. G. White, Phys. Rev., 109:1860 (1958).

44.   T. Piwkowski, Acta Phys. Pol., 15:271 (1956).

45.   G. Schwabe, Z. Naturforsch., 10a:78 (1955).

46.   J. Starkiewicz, L. Sosnowski, and O. Simpson, Nature 158:28 (1946).

47.   R. Ya. Berlaga and L. P. Strakhov, Zh. Tekh. Fiz., 24:943 (1954).

48.   G. Cheroff and S. P. Keller, Phys. Rev., 111:98 (1958).

49.   L. Pensak, Phys. Rev., 109:601, (1958).

50.   B. Goldstein and L. Pensak, J. Appl. Phys., 30:155 (1959).

51.   J. Tauc, Photo and Thermoelectric Effects in Semiconductors, Pergamon Press, Oxford (1962).

52.   J. S. Prener and F. E. Williams, J. Electrochem. Soc., 103:342 (1956).

53.   J. S. Prener and F. E. Williams, Phys. Rev., 101:1427 (1956).

54.   J. S. Prener and F. E. Williams, J. Chem. Phys., 25:361 (1956).

55.   J. S. Prener and F. E. Williams, J. Phys. Radium, 17:667 (1956).

56.   J. S. Prener and D. J. Weil, J. Electrochem. Soc., 106:409 (1959).

57.   W. W. Piper and F. E. Williams, Solid State Phys., 6:96 (1958).

58.   R. H. Bube, Photoconductivity of Solids, Wiley, New York (1960).

59.   L. A. Vinokurov and M. V. Fok, Opt. Spektrosk., 10:225 (1961).

60.   A. F. Kroger, Physica (The Hague), 22:637 (1956).

61.   H. A. Klasens, J. Electrochem. Soc., 100:72 (1953).

62.   Yu. V. Bochkov, A. N. Georgobiani, and G. S. Chilaya, Fiz. Tverd. Tela, 8:1273 (1966).

63.   S. J. Czyzak, D. C. Reynolds, R. C. Allen, and C. C. Reynolds, J. Opt. Soc. Amer., 44:864 (1954).

64.   P. Zalm, Philips Res. Rep., 11:353, 417 (1956).

65.   P. G. Le Comber, W. E. Spear, and A. Weinmann, Brit. J. Appl. Phys., 17:467 (1966).

66.   A. Lempicki, D. R. Frankl, and V. A. Brophy, Phys. Rev., 107:1238 (1957).

67.   V. E. Oranovskii, E. I. Panasyuk, and B. T. Fedyushin, Inzh. Fiz. Zh., 2 (1):39 (1959).

68.   A. N. Georgobiani and H. Friedrich, Fiz. Tverd. Tela (in press); Abstracts of Papers presented at Second All-Union Conf. on $A^{II}B^{VI}$ Compounds, Uzhgorod, 1969 [in Russian], p. 61.

69.   A. Kobayashi and S. Kawaji, J. Chem. Phys., 24:907 (1957).

70.   A. Kobayashi and S. Kawaji, J. Phys. Soc. Jap., 11:369 (1956).

71.   J. Bardeen and S. R. Morrison, Physica (The Hague), 20:873 (1954).

72.   G. Heiland, Z. Phys., 142:415 (1955).

73.   G. Heiland, J. Phys. Chem. Solids, 6:155 (1958).

74.   R. H. Bube, J. Chem. Phys., 27:496 (1957).

75.   D. R. Frankl, Phys. Rev., 100:1105 (1955).

76.   G. F. Alfrey and K. N. R. Taylor, Helv. Phys. Acta, 30:206 (1957).

77.   G. F. Alfrey and I. Cooke, Proc. Phys. Soc., London, B70:1096 (1957).

78.   I. Cooke, J. Chem. Phys., 38:291 (1963).

79.   P. F. Browne, J. Electron., 2:154 (1956).

80.   L. Pauling, The Nature of the Chemical Bond and the Structure of Molecules and Crystals, 3rd ed., Cornell University Press, New York (1960).

81.   A. N. Georgobiani, Tr. Fiz. Inst. Akad. Nauk SSSR, 23:3 (1963) [Soviet Researches on Luminescence, Consultants Bureau, New York (1964), p. 1].

82.   M. V. Fok, Czech. J. Phys., B13:99 (1963).

83.   W. W. Piper, Phys. Rev., 92:23 (1953).

84.   S. Rothschild, Trans. Faraday Soc., 42:635 (1946).

85.   H. C. Froelich, J. Electrochem. Soc., 100:280 (1953).

86.   A. M. Gurvich, Usp. Khim., 35:1495 (1966).
87.   S. Larach and R. E. Shrader, J. Phys. Chem. Solids, 3:159 (1957).
88.   Yu. V. Bochkov, A. N. Georgobiani, N. N. Kisil', L. A. Sysoev, and G. S. Chilaya, Izv. Akad. Nauk SSSR, Ser. Fiz., 30:629 (1966).
89.   A. N. Georgobiani and V. I. Steblin, Fiz. Tekh. Poluprov., 1:934, 956 (1967).
90.   M. Aven and D. A. Cusano, J. Electrochem. Soc., "Program of Pittsburgh Meeting, 1963" (Abstract 26), 110:52C (March 1963).
91.   M. Aven and D. A. Cusano, J. Appl. Phys., 35, 606 (1964).
92.   A. N. Georgobiani and V. I. Steblin, Fiz. Tekh. Poluprov., 1:329 (1967).

ELECTROLUMINESCENCE AND SOME
ELECTRICAL PROPERTIES OF
HOMOGENEOUS ZINC SULFIDE SINGLE CRYSTALS

Yu. V. Bochkov, A. N. Georgobiani, and G. S. Chilaya

Electroluminescence was generated in homogeneous samples of zinc sulfide free of internal contact barriers. The electroluminescence spectrum and dependence of the brightness on the voltage were determined. Ultraviolet electroluminescence, corresponding to interband recombination of electron-hole pairs, was detected and its spectrum investigated. The hypothesis of strong compensation of zinc sulfide was verified. The forbidden band width and the donor level depth were determined from the electrical conductivity of zinc sulfide single crystals. It was found that the bulk electrical conductivity, photoconductivity, and electroluminescence of zinc sulfide were weakly anisotropic.

INTRODUCTION

There is increasing interest in wide-gap semiconductors. This is due to certain problems which have arisen in semiconductor technology. These problems include the development of semiconductor devices capable of operating at high temperatures and high voltages as well as of devices capable of detecting and generating visible and ultraviolet radiation. A semiconductor suitable for use in powerful sources of short-wavelength radiation must not only have a wide forbidden band but also a high probability of radiative transitions. These requirements are satisfied by zinc sulfide, whose thermal forbidden band width is 3.2 ± 0.2 eV [1] and the optical width is 3.7 eV [2]. . Zinc sulfide is the base material of many excellent phosphors. The high efficiency of the luminescence emitted by this compound under various excitation conditions makes it a suitable material for various technical applications.

Zinc sulfide phosphors are currently used as screen coatings in television equipment, luminescent panels, light amplifiers, optrons, and other electroluminescent devices. Thin ZnS films are used as optical filters and selective mirrors. Zinc sulfide is also used in nuclear radiation counters.

However, in spite of the wide applications of zinc sulfide in electroluminescent devices, its electrical properties have not yet been studied sufficiently thoroughly. There are many contradictory and unreliable results. This is because investigations have been carried out on samples in the form of powders, thin films, or small crystallites. The number of investigations carried out on single crystals is relatively small and, in most of them, the authors have failed to satisfy the conditions which must be observed in reliable measurements of the electrical porperties of semiconductors (samples of special and reproducible

configurations, ohmic contacts, elimination of surface currents, etc.). The situation is compli-
cated by the high resistivity of zinc sulfide, which makes it difficult to study its electrical
properties.

The present investigation had a twofold aim: first, a study of the bulk electrical properties of homogeneous crystals of zinc sulfide and, second, a study of the electroluminescence (right up to ultraviolet wavelengths) of these crystals. Electroluminescence may be
excited in two ways: in one case, electron — hole pairs are generated as a result of electrical
breakdown of a crystal or p — n junction subjected to a reverse voltage (this is known as
breakdown luminescence);* in the other case, electroluminescence is due to carrier injection
across homo- or heterojunctions. The first case is interesting because of the presence of
"hot" electrons and holes in the resultant electron — hole plasma. This may give rise to
recombination resulting in the emission of quanta of energies greater than the thermal width
of the forbidden band. Moreover, under these conditions, electroluminescence is emitted by
the whole of that part of a crystal in which the electric field is sufficiently high (this may be
a large part of a crystal, particularly in the case of discharge in a homogeneous crystal).
In the case of p — n junctions, the voltages required to generate electroluminescence are two
or three orders of magnitude lower than those needed for homogeneous crystals. However,
in the presence of a p — n junction only a small part of a crystal ($\sim 10^{-5}$ cm wide) is active
in the emission of electroluminescence and this reduces the integrated radiation power
compared with that which is emitted by homogeneous crystals. This explains, in particular,
why the development of a junction laser has been followed by work on lasers excited with fast
electrons (cathode rays): the lower efficiency in the case of electron bombardment is compensated by the higher output power from a larger active region. In the case of electron
bombardment, the thickness of the active region may reach a few microns. In a high-voltage
electric discharge through a homogeneous crystal, the thickness of the active region can be
at least another order of magnitude larger.

CHAPTER I

REVIEW OF PUBLISHED LITERATURE

§1. Optical and Electrical Properties of Zinc Sulfide

Almost all $A^{II} B^{VI}$ compounds crystallize in such a way that each atom is located at the
center of a regular tetrahedron and atoms of the other element are located at the four corners
of this tetrahedron. Two types of structure can be built up from such tetrahedra: one is the
zinc blende (cubic) structure† and the other is wurtzite (hexagonal) structure. Both these
structures are shown in Fig. 1 [3] . The two polymorphic modifications of zinc sulfide have
the following structure: sulfur atoms form a cubic or hexagonal close-packed structure,
whereas zinc atoms occupy half the tetrahedral vacancies in the close-packed configuration.
The zinc blende structure can be represented as a combination of two close-packed fcc
lattices: one consists of sulfur atoms and the other of zinc atoms. The two lattices penetrate
each other and are shifted along the cube diagonal by a quarter of its length. Wurtzite can

---

* By electric breakdown, we mean here a rapid increase in the electric current as a result of
  the formation of excess carriers but we shall restrict the meaning of electric breakdown to
  a reversible phenomenon which does not damage crystal.
† It is often called also the sphalerite structure.

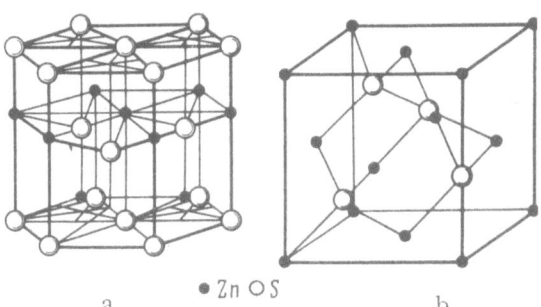

Fig. 1. Hexagonal and cubic modifications of zinc sulfide: a) wurtzite; b) sphalerite.

also be regarded as a combination of two interpenetrating hexagonal close-packed lattices shifted relative to one another along a common threefold axis.

The ordered distribution of atoms in a crystal is established and maintained by forces which are basically electrostatic. Two types of bonding are encountered in semiconductors: ionic and covalent. In the ionic bonding, the positive and negative ions are distributed in an ordered manner between the lattice sites and the principal contribution to the bond energy is made by the Coulomb interaction between these ions. In covalent bonding the valence electrons of atoms forming a molecule are shared and the bond energy consists mainly of the exchange energy of the shared electron pairs. Ionic and covalent bonding may coexist in a given crystal but the relative contributions of the different types of bonding are specific to a given structure.

The type of bonding of a substance is a very important property because the optical and electrical parameters are governed by the type of bonding. Many investigations have been made of the nature of bonds in ZnS. Saksena [4] and Browne [5] found that the ionic type of bonding represented ~ 65% of the total bonding in ZnS. This result was deduced by Saksena from the piezoelectric effect and by Browne from the luminescence efficiency. However, most of the other workers have tended to assume that the bonds in zinc sulfide are mainly covalent. This view is held because the tetrahedral bond configuration in ZnS suggests the predominance of covalent bonding, whereas ionic crystals tend to form structures with large numbers of nearest neighbors (large coordination numbers). This view is supported also by estimates of the ionicity based on the electron affinities of zinc sulfur, i.e., on the energy representing the attraction of bonding electrons to a given atom. An estimate of this kind yielded ~ 30% for the ionic bonding in ZnS (details are given in [6]. Fok [1] compared the energies of the longitudinal and transverse optical phonons and found that the bond ionicity in ZnS was ~ 40%. The predominantly covalent type of bonding is indicated also by the relatively high electron mobility in zinc sulfide, $\mu = 120 \text{ cm}^2 \cdot V^{-1} \cdot \sec^{-1}$ [7], whereas in ionic crystals the mobility is usually a few cgs units ($\text{cm}^2 \cdot V^{-1} \cdot \sec^{-1}$).

There have been several investigations of electrical conduction in ZnS. Until recently, all attempts to prepare p-type crystals have not been successful and it has even been suggested that such crystals cannot be prepared [8]. However, a special treatment has enabled Georgobiani and Steblin [9] to prepare p-type ZnS. Lempicki et al. [10] measured the conductivity of small zinc sulfide crystals. These measurements were carried out on samples in the shape of needles, thin plates, and thin rods. They found that the dark electrical conductivity and photoconductivity are strongly anisotropic. According to Lempicki et al., the dark conductivity along the optic axis is six orders of magnitude lower than at right-angles to this axis. The corresponding ratio of the photoconductivities is ~ 10^4. These values apply to room temperature.

Fok [1] determined the temperature dependence of the electrical conductivity of a zinc sulfide powder in the range 900-1300°K. Fok used this dependence to find the thermal width of the forbidden band of ZnS: $\Delta E_t = 3.2 \pm 0.2$ eV. The temperature dependence of the conductivity was also investigated by Piper [11]. According to Piper, $\Delta E_t = 3.67 \pm 0.1$ eV. However, Fok demonstrated that Piper made an experimental error by measuring the conductivity of the mica insulation rather than of the zinc sulfide itself. Moreover, there are doubts about the agreement between the optical (ΔE_o) and thermal (ΔE_t) widths of a forbidden band, as reported by Piper. This is because the difference between the thermal and optical energy gap is typical of $A^{II} B^{VI}$ compounds. This difference is due to transitions to the polaron state of the electrons and holes formed due to the absorption of light by the crystal lattice. This difference is given by $\delta = \Delta E_o - \Delta E_t = E_n + E_p$, where E_n and E_p are, respectively, the energies of formation of the electron and hole polarons. The energy of polarons usually increases with increasing ionicity of a crystal. In the final analysis this difference between the energy gaps is a consequence of the Franck – Condon principle, which states that optical transitions of electrons are too fast for lattice relaxation. Fok [1] determined the difference between the thermal and optical widths of the forbidden band in zinc sulfide. He found that this difference was 0.55 ± 0.15 eV. The values of this difference are known also for some other $A^{II} B^{VI}$ compounds. For example, $\delta \approx 0.12$ eV for CdTe, $\delta \approx 0.14$ eV for CdS, $\delta \approx 0.28$ eV for CdSe [12].

Very extensive information on the structure of a solid can be obtained from optical investigations. There have been many studies of the optical properties of zinc sulfide in the fundamental absorption region. However, sufficiently comprehensive experimental data, which would make it possible to draw undisputed conclusions, are not yet available. One of the most pressing and contentious problems is the nature of discrete levels responsible for the strong absorption in the $\lambda = 300-330$ nm range. This absorption is probably due to the overlap of the fundamental and impurity absorption regions. However, many authors are of the opinion that the absorption in this range of wavelengths is due to band–band transitions. Thus, for example, Alentsev and Panasyuk [13] found that the absorption coefficient was high (5×10^4 cm^{-1}) at $\lambda = 312$ nm (3.97 eV). This large value of the coefficient could not be due to the absorption by impurities and, therefore, it was attributed by Fok [1] to band–band transitions. Gross et al. [14] investigated low-temperature absorption in the same range of wavelengths and found narrow exciton bands.

Cardona and Harbeke [2] investigated the reflection spectra of zinc sulfide in a wide range of wavelengths extending beyond the fundamental absorption edge. Figure 2 shows their spectra of polarized light reflected by hexagonal and cubic crystals of ZnS. The long-wavelength peaks were found to be associated with the fundamental absorption edge. The doublet nature of the edge peak of cubic ZnS (3.66 and 3.76 eV) were attributed to the spin–orbit splitting of the valence band. (The thermoreflection of cubic AnS was investigated recently [15]. The peaks of the edge doublet were found to lie at 3.690 and 3.762 eV.) Table 1, based on the results of Cardona and Harbeke [2], gives the positions of the reflection maxima R and the corresponding calculated positions of the absorption minima k. The last three columns of Table 1 give the transitions in the momentum space which correspond to these maxima and minima. The data in this table were used by Cohen and Bergstresser [16] to calculate the (Fig. 3). We can easily see that the extrema of the conduction (Γ_1) and valence (Γ_{15}) bands are

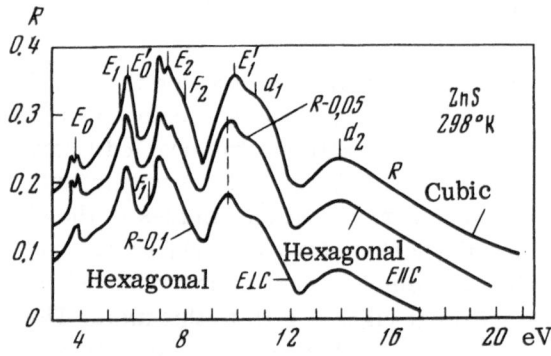

Fig. 2. Reflection spectra of hexagonal and cubic zinc sulfide.

TABLE 1. Energies (in eV) of Absorption (k) and Reflection (R) Peaks and Associated Transitions in the Brillouin Zone of ZnS

Reflection peak designation	Type of peak	Cubic crystals	Hexagonal crystals		Cubic crystals	Hexagonal crystals	
			E ∥ C	E ⊥ C		E ∥ C	E ⊥ C
E_0	R	3.66	3.74	3.78	$\Gamma_{15} \to \Gamma_1$	$\Gamma_1 \to \Gamma_1$	$\left.\begin{matrix}\Gamma_5\\\Gamma_1\end{matrix}\right\} \to \Gamma_1$
		3.76	3.88	3.87			
E_0'	k	5.82	5.86	5.80	$\Gamma_{15} \to \Gamma_{15}$	$\Gamma_5 \to \Gamma_1$	$\Gamma_5 \to \Gamma_1$
	R	5.79	5.76	5.74		$\Gamma_1 \to \Gamma_1$	$\Gamma_1 \to \Gamma_5$
						Forbidden	$\Delta_5 - \Delta_1$
A	k	5.3					
	R	5.4		5.5			
E_1					$\Lambda_3 \to \Lambda_1$		
B	k		5.6			$M_1 \to M_1;$	$M_2 \to M_4;$
	R					$K_2 \to K_2;$	$M_1 \to M_4;$
						$\Sigma_1 \to \Sigma_2$	$\Sigma_1 \to \Sigma_2$
E_1'	k	9.65	9.56	9.43	$L_3 \to L_3$	$\Gamma_6 \to \Gamma_6$ etc.	Unknown
	R	9.78	9.73	9.61			
E_2	k	7.03; 7.35	7.01; 7.50	7.00; 7.5	$X_5 \to \begin{cases}X_4\\X_3\end{cases}$	Unknown	"
	R	6.99; 7.41	6.98; 7.56	7.00; 7.52			
d_1	k	10.6			$L_3 \to L_1$	"	$\Gamma_6 \to \Gamma_3$ etc.
	R	10.8	10.8	10.8			
d_2	k	14.3			From d electrons	From d electrons	From d electrons
	R	13.8	13.8	13.8			
F_1	k			6.6	Unknown	Forbidden	Unknown
	R			6.6			
F_2	k	7.9				Unknown	"
	R	7.9			"		

located at the point k whose coordinates are (0, 0, 0) at the point Γ. The occurrence of direct transitions in ZnS is predicted also by Birman [17].

Shalimova and Morozova [19] and Arapova [20] investigated absorption in thin sublimated films of ZnS. They found strong bands in the 200–250 nm range. These bands probably represented the energy band structure determined by Cardona or Harbeke [2] or the exciton states of higher bands.

Much information on the energy structure of a semiconductor can be obtained also from its luminescence spectrum. Since the beginning of this century it has been known that the luminuecence of ZnS is associated with the presence of few impurities which form luminescence centers. The impurities responsible for the luminescence are known as activators (the introduction of such impurities into the lattice is called activation in papers on luminescence and doping in papers on semiconductors). However, it has been found that other impurities, known as coactivators, are necessary if good phosphors are needed. Many authors are now of the opinion that the luminescence centers are formed from activators and coactivators.

The positions of activators and coactivators of $A^{II}B^{VI}$ crystals is shown as part of the periodic table in Fig. 4 [21]. Figure 5 shows the positions of the maxima in the luminescence spectra of cubic ZnS crystals containing various impurities. The results given in Fig. 5 can be found in Rebane's book [22]. The same bands are usually observed also in hexagonal zinc sulfide.

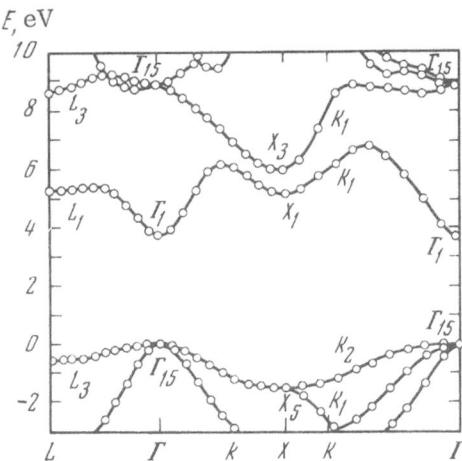

Fig. 3. Energy band structure of the cubic modification of zinc sulfide.

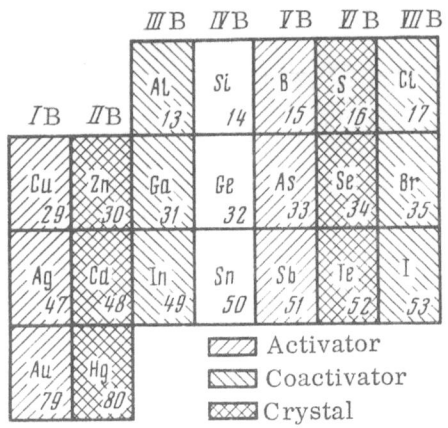

Fig. 4. Positions of activators and coactivators, suitable for $A^{II}B^{VI}$ compounds, in Mendeleev's periodic table.

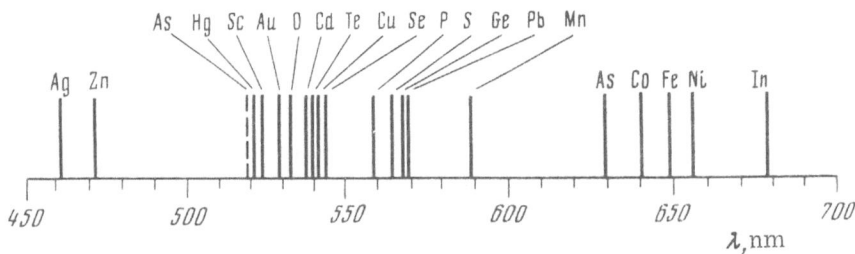

Fig. 5. Schematic representation of the positions of the maxima in the luminescence spectra of cubic zinc sulfide containing various impurities.

It is worth mentioning that activators are usually acceptors, whereas coactivators are normally donors. The replacement of zinc with an element of group I or of sulfur with an element of group V produces a defect which tends to capture one electron so as to complete its outer shell. This gives rise to discrete acceptor levels above the valence band. If zinc is replaced by an element of group III or sulfur by an element of group VII we find that there is one "spare" electron. This gives rise to discrete donor levels below the bottom of the conduction band. The simultaneous introduction of an activator and a coactivator results in a transfer of an electron from a donor to an acceptor, i.e., it results in compensation.

It is nowadays generally accepted that zinc sulfide is a strongly compensated semiconductor. This conclusion rests on the observation that a large amount of an activator can be introduced only if a crystal contains approximately the same amount of a coactivator.

According to some authors, a coactivator is needed not only to compensate the charge of a crystal as a whole but also to form luminescence centers in combination with an activator. The idea that compensating impurities combine to form such centers was first put forward by Prener and Williams [23-25]. According to their theory, the formation of a donor—acceptor pair occurs in the following way: an activator and a coactivator become associated in the lattice, i.e., they occupy nearby positions. They may lie very close to one another (forming first-order pairs) or they may be separated by some distance (higher-order pairs). The formation of such acceptor—donor pairs results in the overlap of the wave functions of the acceptor and donor so that an electron may be transferred from a donor to an acceptor level.

According to Prener and Williams, the first-order pairs are responsible for the short-wavelength luminescence and the higher-order pairs are responsible for the long-wavelength bands.

The luminescence associated with defects is observed also in zinc sulfide crystals which have not been doped (at least not deliberately). Such luminescence is called self-activated. The origin of the self-activated luminescence has been the subject of many theories and hypotheses. Two hypotheses are current popular: some authors attribute the self-activated luminescence to zinc vacancies [26] and others attribute it to complexes formed from zinc vacancies and group VII elements which replace sulfur [27, 28].

A zinc vacancy is short of two electrons (out of eight) in the four bonds which it forms with the neighboring sulfur atoms, i.e., there are two holes in the vicinity of such a vacancy. Thus, zinc vacancies are acceptors. Similarly, sulfur vacancies form defects, each of which is associated with two spare electrons. This means that sulfur vacancies act as donors.

This section gives a brief summary of the available information on the electrical and optical properties of ZnS. These properties include also electroluminescence. However, since the present paper is concerned primarily with electroluminescence, this subject deserves a special section.

§2. Electroluminescence of Zinc Sulfide

Electroluminescence is that type of luminescence in which a phosphor acquires the necessary energy directly from an electric field. Hundreds of papers have been written about the electroluminescence of zinc sulfide. Most of the investigations published so far have been carried out on fine ZnS powders. A powder of $\sim 10\,\mu$ grain size is usually made up into an electroluminescent cell, which is a capacitor filled with zinc sulfide suspended in a liquid or polymerized dielectric. However, much work has been done also on needle-shaped, wedge-shaped, or plate-like crystals grown from the vapor phase.

The activator in electroluminescent phosphors based on ZnS is usually copper by itself or copper and manganese [7, 29-32]. Some electroluminescent ZnS crystals have been activated with phosphorus [33] or silver [33, 34]. The coactivators are the same as in photoluminescent phosphors. Since electroluminescent phosphors emit the same luminescence bands as the corresponding photoluminescent crystals, it is usual to assume that the same luminescence centers are formed in both types of phosphor. However, good-quality electroluminescent phosphors usually need a higher doping level than photoluminescent crystals. This normally results in the precipitation of a high-conductivity phase (for example, Cu_2S) on the surface of a crystal or a grain. This phase is known to play an important role in electroluminescence (this point will be discussed in greater detail in the present paper).

Two types of electroluminescence can be distinguished on the basis of the excitation process:

1. Low-voltage (injection) electroluminescence. In this case, electron-hole pairs are generated by the injection of minority carriers into a semiconductor through a contact with a metal or another semiconductor. The excitation of such luminescence usually does not require high values of the voltage. This luminescence is observed, for example, in forward-biased p — n junctions.

2. High-voltage (breakdown) electroluminescence. In this case, the formation of electron — hole pairs and the excitation of luminescence centers are the result of processes occurring during electric breakdown of a solid (field and impact ionization).

Since the present paper is concerned with the high-voltage electroluminescence, we must consider in greater detail the processes which result in electric breakdown of a semiconductor.

A. Direct Ionization by an Electric Field (Tunnel Effect). A strong electric field may generate nonequilibrium electrons and holes in a semiconductor. The crystal lattice and local centers are ionized by the tunnel leakage of electrons across potential barriers. Such tunnel transitions become possible because of the band bending in a strong electric field. The mechanism resulting in the transfer of electrons to the conduction band as a result of their tunnel leakage across potential barriers has been considered first by Zener [35] and, therefore, the phenomenon is known as the Zener effect. The probability that an electron reaches the conduction band, calculated per unit time and per unit volume, is given by the formula

$$W_{tu} = \frac{eEd}{2\pi\hbar} \exp\left\{-\frac{\pi}{4e\hbar E}\sqrt{2m^*}\Delta^{3/2}\right\},$$ (I.1)

where E is the electric field; e is the electronic charge; m^* is the effective mass of an electron; Δ is the forbidden band width; d is the lattice period.

Equation (I.1) applies only to direct transitions. Indirect transitions have been considered by Keldysh [36]. The probability of ionization per unit time increases when the temperature is raised because the number of phonons increases. In the temperature range

$$T_0 < T < T_{D},$$

where T_D is the Debye temperature and $T_0 \approx 10^{-4}E$ (here E is the field in V/cm), the formula for the tunneling probability W_{tu}^0 is of the form

$$W_{tu}^0 = a_0 \frac{1}{\exp\left(\frac{\hbar\omega_q}{kT}\right) - 1} \exp\left[-\frac{4\sqrt{2m^*}(\Delta \pm \hbar\omega_q)^{3/2}}{3\hbar eE}\right],$$ (I.2)

where a_0 is a quantity which depends weakly on the field and temperature; $\hbar\omega_q$ is the energy of a phonon required to satisfy the law of conservation of momentum; k is the Boltzmann constant; the plus and minus signs are used depending on whether a transition involves the emission or the absorption of a phonon. The probability of field ionization increases when the temperature is increased also because the forbidden band becomes narrower at high temperatures. The relative narrowing of the forbidden band width is of the order of the linear expansion coefficient of the crystal lattice. The narrowing of the forbidden band in zinc sulfide has been investigated by many workers [37-40] and they have found that the temperature coefficient of the forbidden band width $d\Delta/dT$ ranges from -2×10^{-4} to -9×10^{-4} eV/deg.

Phonons can make a considerable energy contribution to the tunnel ionization if the temperature is high and the field is weak, i.e., when the relationship $E < (kT/ed)$ is satisfied. These tunnel transitions are assisted by many phonons and have been considered theoretically by Keldysh [36]. The probability of such a transition is given by the formula

$$W_{tu}^T = a \exp\left[-\frac{\Delta}{kT} + \frac{1}{24m^*}\frac{(e\hbar E)^2}{(kT)^3}\right],$$ (I.3)

where a is a quantity which depends weakly on the temperature. In some ways, the above formula is intermediate between the thermal ionization equation and Eqs. (I.1), (I.2).

If we consider the ionization of a local center with a level close to the band edge, we must replace the electronic charge e in the above formulas with the expression

$$e^* = \frac{e}{\sqrt{\varepsilon_\infty}},$$ (I.4)

where ε_∞ is the high-frequency permittivity.

B. Impact Ionization. The appearance of nonequilibrium carriers in a strong electric field need not be due to the tunnel effect but may be caused by the avalanche multiplication of carriers known as impact ionization. Such ionization occurs as follows. When a strong electric field is applied to a crystal a conduction-band electron or a free hole acquires an energy sufficient to knock out an electron from the valence band and transfer it to the conduction band. The knocked-out electron is replaced in the valence band by a new hole. Thus, instead of one free carrier we now have three. All three carriers acquire additional energy in the electric field and the multiplication process is repeated. This gives rise to an avalanche which may end in electric breakdown. Impact ionization of impurity centers does not differ basically from the process in which isolated atoms are involved.

Free carriers are scattered by defects (impurities, vacancies, phonons, etc.). Such defects make it difficult for an electron to acquire an energy necessary for impact ionization. The lattice vibrations in which neighboring ions or atoms oscillate in antiphase (optical phonons) have a particularly strong influence on the motion of carriers. This is because such vibrations are more inhomogeneous than the others. The frequency of interaction between electrons and optical phonons depends strongly on the ionicity of a crystal. This is because oppositely charged ions alternate in an ionic crystal and this gives rise to strong perturbing fields. Since the interaction of electrons with the lattice is particularly strong in ionic crystals, such electrons are in thermal equilibrium with the lattice even if the field is strong. According to Callen [41], the average energy of electrons in prebreakdown fields amounts only to a few kT. Thus, only a very small fraction of electrons can acquire energies sufficient for impact ionization. Electrons lose their energy mainly by heating the lattice. Consequently, thermal breakdown in such crystals may occur in fields which are insufficient to produce significant electroluminescence. The situation is quite difficult in covalent crystals: in this case, the frequency of electron — phonon collisions is several orders of magnitude less than in ionic crystals. Consequently, a considerable fraction of electrons acquires an energy considerably greater than the average thermal energy. Therefore, impact ionization processes in covalent crystals begin at much lower fields than in ionic crystals.

The average probability of impact ionization in a given electric field is governed by the probability that an electron acquires an energy ε_i sufficient for the ionization of the crystal lattice or of impurities. In the lattice ionization the value of ε_i is always greater than the forbidden band width Δ. This is a consequence of the law of conservation of momentum. For spherically symmetrical energy bands we have $\varepsilon_i = (3/2)\Delta$. The probability of impact ionization is usually calculated for two extreme cases. In one case, an electron traverses without collisions a distance sufficient to acquire the ionization energy (this is possible because of fluctuations in the mean free path). The case is analogous to the acceleration of electrons in gaseous discharges, which was considered first by Townsend. Consequently, similar phenomena in semiconductors are often called Townsend discharges. According to Townsend, the probability of impact ionization depends exponentially on the field:

$$W_{\text{im}}^{0} \propto \exp\left(-\frac{\text{const}}{E}\right). \tag{I. 5}$$

The above formula describes the first (collisionless) case and applies to relatively weak fields. In the second case, the field is assumed to be stronger and an important role is played by those electrons which acquire the ionization energy in several stages separated by collisions. The important point is that an electron should acquire, in each free path, an energy greater than that which it loses in the next collision. The acquisition of energy by the electron has the characteristics of diffusion and, therefore, this mechanism is known as the diffusion case.

A rigorous calculation of the dependence of the probability of impact ionization on the field requires a solution of the corresponding transport equation. Such a solution for the diffusion approximation was obtained by Chuenkov [42] and Keldysh [43]. The probability of impact ionization in the diffusion approximation is given by a formula which differs somewhat from the Townsend equation:

$$W_{im}^{d} \propto \exp\left(-\frac{\text{const}}{E^{\perp}}\right). \qquad (I.6)$$

At a later stage the problem was solved for the general case: Keldysh obtained the solution for covalent crystals [44] and Chuenkov for ionic solids [45]. The general dependence of the impact ionization probability on the field is given by

$$W_{im}^{gen} \propto \exp\left(-\frac{\text{const}}{E^{\gamma(E)}}\right). \qquad (I.7)$$

The value of γ for covalent crystals is close to 1 if the field is weak and if $\hbar\omega_0 \gtrless eEl > kT$. If the field is strong, γ is close to 2. For ionic crystals we have $\gamma = 2$ if $eEl \ (\hbar\omega_0) > \hbar\omega_0 > kT$ or $eEl \ (kT) > KT > \hbar\omega_0$; if these inequalities are not satisfied we find that $2 > \gamma > 1$, where $\hbar\omega_0$ is the energy of optical phonons and l is the mean free path of electrons.

If we know the probability of impact ionization we can calculate the impact ionization coefficient $\chi_{im}(E, T)$, i.e., the number of ionizations caused by an electron moving in a field E (this number is taken per unit length):

$$\varkappa_{im} = \frac{W_{im}}{u}, \qquad (I.8)$$

where u is the drift velocity of electrons. If we ignore special cases (such as very thin films) we find that impact ionization occurs in fields much lower than those required for the tunnel effect.

The processes considered here result in electric breakdown and, therefore, in breakdown electroluminescence which is significant in fields $\gtrsim 10^5$ V/cm. However, the electroluminescence of ZnS is frequently obtained in much weaker average fields of the order of 10^4 V/cm. This is due to a nonuniform distribution of the field in phosphor crystals because of the formation of depletion-type Mott—Schottky barriers in which the electric field is concentrated. The relationship between the maximum electric field E in such a barrier and the applied voltage is determined by the space—charge distribution in the barrier. If the charge is distributed uniformly we find that $E \propto \sqrt{V}$. It is assumed that the field in such barriers is sufficient to generate electroluminescence. This is confirmed by the absence of the influence of strong magnetic fields (up to 1.7×10^5 Oe) on electroluminescence [46]. These values of the magnetic field can be used to estimate the electric field in the barriers which exceeds 10^5 V/cm. Chukova [47] has established that an electric current is rectified quite strongly by an electroluminescent capacitor and that Mott—Schottky barriers play an important role in the rectification process. A study of the rectification effect enabled Chukova to estimate the field in the barriers in electroluminescent crystals. This field is of the order of 10^6 V/cm.

The nonuniform distribution of the electric field in crystals is responsible for the nonuniform distribution of their electroluminescence. Microscopic studies of the electroluminescence of zinc sulfide single crystals have established that the luminescence is concentrated at isolated luminous points and lines [48, 49]. Lehmann [50] used a microscope to examine the electroluminescence of separate grains of a phosphor mixed with a dielectric whose refractive index was the same as that of ZnS, i.e., the dielectric was used as an immersion medium. Lehmann found that the luminescence was concentrated at isolated points and that these points were located at the edges of opaque regions occupied by a precipitated second phase. We have mentioned earlier that high-quality electroluminescent phosphors contain a large amount of copper so that a second phase in the form of Cu_2S is precipitated on surfaces or at dislocations. Several workers have established that the presence of this second phase is an important factor in the electroluminescence of ZnS [7, 51, 52]. Zinc sulfide and Cu_2S form a p—n heterojunction structure. The application of a reverse voltage to such a structure results in the concentration of the electric field in the junction. Some authors are of the opinion that the second phase is essential for the concentration of the field in the junction barrier. However, it is now known that depletion barriers in which the electric field is concentrated can form also on the surface of a semiconducting crystal because of the loss of free carriers. Therefore, Georgobiani and Fok [53] , who investigated the mechanism of electroluminescence of ZnS powders, have suggested that an electric field results in tunnel transitions of electrons from the valence band of Cu_2S to the conduction band of zinc sulfide (the existence of such transitions has been demonstrated for tunnel diodes). Subsequently, these electrons are accelerated to energies sufficient for impact ionization of the crystal lattice and of the luminescence centers. Georgobiani and Fok have also concluded that these tunnel transitions govern the experimentally observed dependence of the electroluminescence brightness on the applied voltage [7, 54]:

$$B = B_0 \exp\left[-\frac{b}{\sqrt{U}}\right], \qquad\qquad (I. 9)$$

where B is the brightness; U is the applied voltage; B_0 and b are quantities which are independent of the voltage.

The energy band scheme of a heterojunction in a zinc sulfide phosphor is shown in Fig. 6. The arrows in Fig. 6b indicate tunnel transitions. The scheme is plotted on the assumption that ZnS has n-type conduction and Cu_2S has p-type conduction.

This mechanism of electroluminescence was tested by investigating specially prepared $ZnS:Cu_2S$ heterojunctions [55]. A study of the electroluminescence of such heterojunctions, subjected to a reverse bias, demonstrated that their characteristics were similar to those of the electroluminescence of zinc sulfide phosphors. Since similar characteristics have been

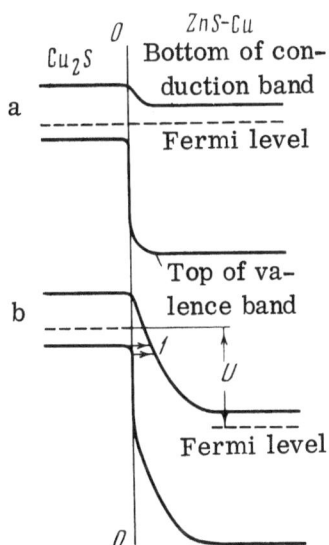

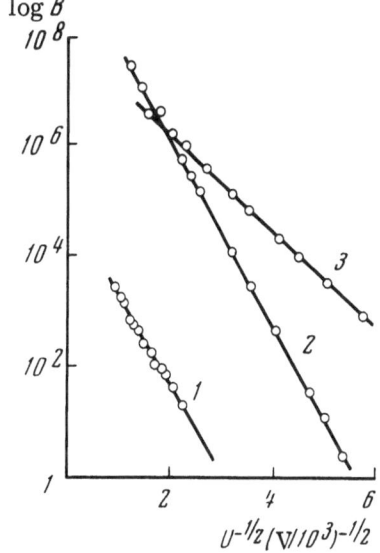

Fig. 6. Energy band scheme of a ZnS:Cu phosphor in contact with Cu_2S: a) in the absence of an external voltage; b) under a voltage U.

Fig. 7. Dependence of the logarithm of the brightness on $U^{-\frac{1}{2}}$; 1) $ZnS:Li_2S$; 2) $ZnS:Cu:Mn$; 3) $ZnS:CU:Cl$.

observed also for electroluminescent single crystals grown from the vapor phase, it is natural to assume that heterojunctions of this type occur also in single crystals and that luminous points and lines lie close to these junctions.

In the preceding section we have mentioned the strong anisotropy of the electrical conductivity and photoconductivity of the crystals grown from the vapor phase. A similar strong anisotropy is observed in the electroluminescence of these crystals. However, there is no agreement on the nature of this anisotropy. According to Lempicki et al. /10/, the electroluminescence is three orders of magnitude stronger at right angles to the optic axis than the electroluminescence parallel to this axis. Moreover, it is reported in [48, 49] that the electroluminescence is completely absent when the field is applied along the C axis. Diametrically opposite results are reported in [56], i.e., the electroluminescence is found only when the field is applied along the C (optic) axis.

We have mentioned earlier that the dependence of the average electroluminescence brightness on the voltage is described satisfactorily by Eq. (I.9) in the case of powder phosphors and crystals grown from the vapor phase [7, 53, 54]. Dependences of this type are shown in Fig. 7.

In that part of a crystal where the electric field is concentrated as a result of the formation of Mott—Schottky barriers,* the field is given by $E \propto \sqrt{V}$. This relationship between the field and voltage is obtained in all cases when the space charge density in the barrier is uniform. This field-voltage relationship can be used to rewrite Eq. (I. 9) in the form

$$B = B_0 \exp\left[-\frac{b_1}{E}\right].$$

(I. 10)

This dependence of the brightness on the field is usually obtained from the formulas for the tunnel and impact ionization probabilities.

A somewhat different relationship has been reported for thin sublimated films [57]:

$$B = a \exp\left[b\sqrt{E}\right],$$

(I. 11)

but in all cases the dependence of the brightness on the voltage is very strong, which is a typical feature of the processes accompanying electric breakdown.

We have considered the electroluminescence of zinc sulfide in detail because this compound is the subject of our paper. It must be stressed that electroluminescence is emitted by a very large number of elemental and compound semiconductors. Most of the investigations have been carried out on powders, thin films, or p—n junctions. In the last case, the studies have been concerned with injection electroluminescence. There have been very few investigations of the breakdown electroluminescence in homogeneous crystals. Such electroluminescence has been discovered in a great variety of crystals ranging from materials with very narrow forbidden bands (indium antimonide [58]) and ending with ionic crystals. In particular, electroluminescence of KI:Cu, KBr:Cu, KCl:Cu crystals was reported at the International Conference on Luminescence held in 1966 in Budapest [59]. These crystals were of the special shape used in investigations of electric breakdown in dielectrics: the central part of a sample was reduced to a thickness of 10-100 μ.† This made it possible to generate impurity electroluminescence of these compounds in average fields of $\sim 10^5$-10^6 V/cm. This value should be

---

* Such barriers can form at nonohmic contacts, in p—n homo- and heterojunctions, and at the edges of an insulated crystal (during the period when the field is zero).

†Such a very thin central region was achieved by drilling from two opposite sides. The rest of the sample prevented surface breakdown and breakdown in air.

compared with 10^2 V/cm, which was all that was required to excite intrinsic electroluminescence in InSb.

The excitation of electroluminescence in some substances is made easier by the formation of domains in which the electric field is concentrated. Thus, for example, CdS emits electroluminescence right up to photon energies very close to the forbidden band width even in very low fields $\sim 4 \times 10^2$ V/cm [60]. Theoretical calculations indicate that, in the absence of domains, one would require fields two or three orders of magnitude higher.

CHAPTER II

EXPERIMENTAL METHOD

In the preceding chapter we have mentioned that the electroluminescence of ZnS and its electrical properties are usually investigated without satisfying all the conditions which ensure that reliable results are obtained. In the case of semiconductors these conditions are: elimination of contact phenomena by the formation of ohmic contacts; elimination of surface currents; preparation of sufficiently homogeneous samples with a specified configuration and crystallographic orientation. This failure to satisfy the necessary conditions has been due to the lack of well-formed large crystals of zinc sulfide. Measurements have been carried out on powders of 1-10 μ grain size or on needle-shaped, wedge-shaped, or plate-like crystals. Larger samples have been used but these are normally inhomogeneous. Such crystals are inhomogeneous because they are usually prepared from the vapor phase at 1050-1250°C and these temperatures overlap the range of phase transitions (950-1320°C) [61, 62].

It has recently become possible to grow large zinc sulfide crystals from melts. The crystallization temperature in such a growth is much higher than the phase transition temperatures. This has made it possible to investigate the bulk electrical properties of homogeneous zinc sulfides. Such an investigation is the subject of the present paper.

It is known that the electrical and optical properties of a semiconductor are governed basically by the spectrum of its electron energy states and by the distribution of electrons over these states. The data on such states are normally obtained from the absorption and reflection spectra (the optical width of the forbidden band and the optical depth of levels), from the temperature dependence of the electrical conductivity (the thermal width of the forbidden band and the thermal depth of the levels), or from the luminescence spectra. We have mentioned that the optical and thermal widths of the forbidden band differ considerably for substances with appreciable ionicity (which include zinc sulfide). This difference is associated with the transitions of electrons and holes to the polaron state. Since the recombination radiation (luminescence) spectrum depends on the energy states of electrons and holes from which transitions begin, it follows that such a spectrum may include quanta corresponding to the thermal width of the forbidden band and the thermal depth of local levels, whereas quanta of higher energies may be emitted in transitions of "hot" electrons and holes. The absorption spectra are normally investigated using thin crystals or sublimated films $\lesssim 1~\mu$ thick. It is difficult to prepare very thin samples from large crystals. Therefore, we deduced the electron states from the temperature dependence of the electrical conductivity and from the electroluminescence spectra. Since we were initially unable to detect electroluminescence in crystals grown from the melt,* the generation of electroluminescence and a study of its characteristics was our first task.

---

* This was evidently because we employed the standard technique for the excitation of electroluminescence which failed to prevent thermal breakdown in large crystals.

§1. Preparation of Samples

We investigated zinc sulfide single crystals grown at the All-Union Scientific-Research Institute of Single Crystals under the direction of L. A. Sysoev. Single crystals were grown from the melt under an inert-gas pressure by the Stockbarger method. Large crystals of up to 30 mm in diameter and up to 100 mm high grew in a furnace under an argon pressure of 100 atm. The rate of growth of these crystals was 18 mm/h. The crystals were grown on seeds oriented along the C axis. The raw material was a zinc sulfide powder of the phosphor grade, purified so as to remove oxides. Crystallization took place at 1850°C.

A characteristic feature of these crystals was a strong tendency to growth along certain planes. This enabled us to prepare large, plane-parallel plates as well as four- and six-sided prisms.

A prism of this type is shown in Fig. 8. The prism is externally homogeneous and the same is true of the microscopic patterns obtained by etching (to reveal dislocations). These patterns show only tetrahedral etch-pits in the cleavage planes and no hexagonal pits. At right angles to these planes only the hexagonal pits can be seen. The direction parallel to the cleavage planes is also parallel to the C axis.

The crystals were oriented (the C axis was determined) by cleaving or by examination in a polarzing microscope. Crystals oriented in this way were used to prepare samples of the required configuration. This was done on a specially constructed cutting lathe (with a diamond or corundum disk) and a UM1-04 lathe intended for the ultrasonic machining of crystals. The sides other than the cleavage faces were ground and polished using fine-grained chromium oxide.

An etchant consisting of three parts of H_2SO_4, one part of H_2O_2, and one part of water was tried. It was known that this etchant produced satisfactory results in the case of cadmium sulfide. However, the best results for zinc sulfide were obtained using a specially prepared etchant consisting of an aqueous solution of nitric acid and chromium trioxide (one part of HNO_3, two parts of CrO_3, and three parts of water). This etchant worked best when hot. Satisfactory cleaning was achieved by etching for 5 min at 80°C and washing in distilled water.

An ohmic contact with a semiconductor of a given type of conduction should consist of a metal which produces the same type of conduction in this semiconductor. We have mentioned in the preceding chapter that grown zinc sulfide crystals have n-type conduction and only a

Fig. 8. Six-sided prism cleaved from a ZnS single crystal (natural size).

special and fairly complex treatment produces p-type crystals. Therefore, Cooke [63] produced ohmic contacts by evaporating group II elements (In, Ga, Al) in 10^{-6} mm Hg vacuum (these elements were known to form donor centers in ZnS). This was followed by thermal diffusion in the same vacuum at 800°C until the contact barrier was destroyed.

Whenever the value of the electrical conductivity was such that we were able to measure the thermoelectric power, we found that this power indicated n-type conduction. Ohmic contacts were prepared by the Cooke method. The contact material was spectroscopically pure indium. This material was evaporated from a tungsten boat in a vacuum chamber kept at $\sim 10^{-6}$ mm Hg. Sample holders were made of spectroscopically pure graphite, cleaned by boiling in distilled water, and then outgassed. The evaporation indium was melted to form a contact by heating for 10 min at about 600°C.

We found later that similar results were achieved by heating indium in a hydrogen stream passed through a hermetically sealed chamber. The hydrogen was generated in a Kipp unit by reaction between zinc and hydrochloric acid. The hydrogen was bubbled through water before admission to the chamber. A crystal was heated by a tungsten spiral via a quartz substrate. A small piece of indium was placed on the crystal. The chamber had a viewing part through which the melting of the indium could be observed by means of a binocular microscope of the MBS-2 type. The best results were obtained by adopting the following procedure. An electric current was passed through the heater until the indium melted and spread on the surface of the crystal. At that moment the heater was switched off quickly and the whole chamber was cooled by a fan. The hydrogen supply was stopped when the sample cooled completely.

§2. Apparatus

A light-tight thermostat, shown schematically in Fig. 12 of the preceding article, was used to investigate the electrical conductivity. The sample under investigation was clamped between platinum electrodes and the leads of these electrodes passed through quartz holders.

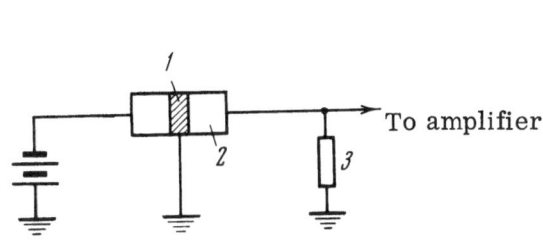

Fig. 9. Circuit for measuring the temperature dependence of the conductivity: 1) guard ring; 2) sample; 3) standard resistor.

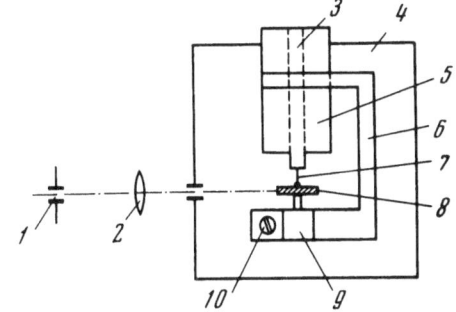

Fig. 10. Sample holder for investigating electroluminescence of ZnS single crystals: 1) entry slit of a monochromator; 2) lens; 3) waveguide; 4) Plexiglas chamber movable along three directions; 5) waveguide; 6) bracket; 7) copper electrode; 8) sample; 9) copper base for heat removal; 10 screw.

A third platinum electrode was pressed against the guard ring and this electrode passed through a quartz tube. A copper—constantan thermocouple was employed in the temperature measurements. The whole assembly was mounted on an asbestos base and covered with a nickel screen and a quartz envelope which was surrounded by heater windings. The thermostat was constructed in such a way that measurements up to ~800°C could be carried out. The thermostat was made light-tight by metal screens which also provided electromagnetic screening.

A dc electrometer amplifier of the U1-2 type was used to measure the current. A 70 V battery was used as the voltage source. A block diagram of the circuit used in the measurements of the electrical conductivity is given in Fig. 9. The sensitivity limit in the measurements of the electrical conductivity was $\sim 10^{-14}$ $\Omega^{-1} \cdot cm^{-1}$. The accuracy of the absolute measurements at the sensitivity limit was 10% and that of the relative measurements was 7%.

The sign of the thermoelectric power was determined using the same apparatus as that employed in measurements of the electrical conductivity. The U1-2 amplifier was switched to the emf-measuring mode. The sensitivity limit in the measurements of the emf was ± 10 mV and the accuracy was $\pm 5\%$. The same thermostat was used but the upper electrode was replaced with a thermoelectric probe. A heater was wound onto the quartz holder of this probe. This heater was switched off during the actual measurements to avoid generating stray emf's.

The electroluminiescence spectra were recorded with a photometric unit based on an SPM-1 Zeiss monochromator (operating in the wavelength range 200-1000 nm) and a quartz photomultiplier of the FÉU-39 type (sensitivity range 160-600 nm) was used as a light detector.

A quartz lens was used to focus electroluminescent light onto the entry slit of the monochromator (Fig. 10). The sample itself was placed in a specially constructed chamber which protected the experimenter from high voltages. The chamber had a mechanism which made it possible to rotate the chamber or to move it along three mutually perpendicular directions. An additional mechanism allowed us to rotate the sample about the vertical axis of the chamber. This made it possible to focus a sample subjected to high voltage. One of the electrodes of the sample was pressed against a copper base mounted at the end of a coaxial waveguide consisting of two stainless steel tubes (the gap between the tube walls was 25 nm). The wave impedance of the waveguide was matched to the output impedance of a pulse generator. A microscope with a long-focus objective was adapted for microscopic investigations in this chamber.

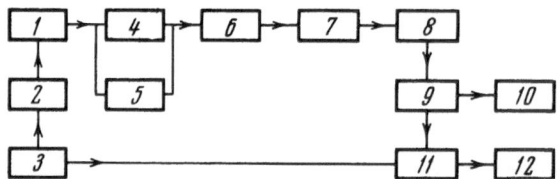

Fig. 11. Block diagram of apparatus for investigating electroluminescence of ZnS crystals: 1) high-voltage pulse generator; 2), 3) GI-3M and MGI-2 pulse generators; 4) ZnS crystal; 5) VLI-2 voltmeter; 6) SPM-1 monochromator; 7) FÉU-39 photomultiplier; 8) cathode follower; 9) USh-2 wide band amplifier; 10) DÉSO-1 oscillograph; 11) synchronous detector; 12) ÉPP-09 automatic recorder.

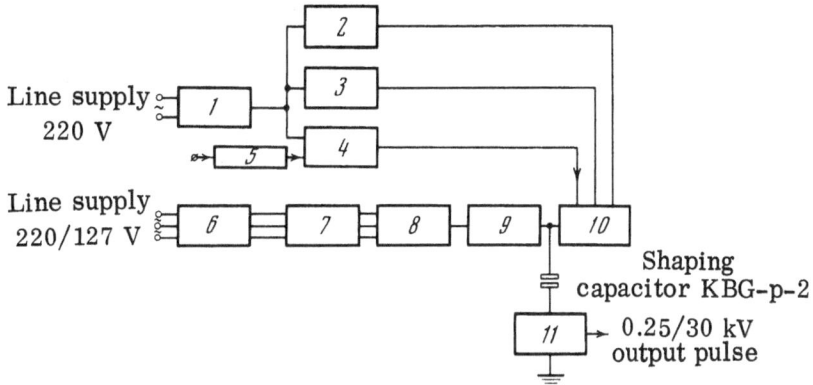

Fig. 12. Functional circuit of the high-voltage pulse generator:
1) S-0.75 voltage stabilizer; 2) screen-grid rectifier; 3) bias rectifier; 4) triggering pulse stage; 5) external stabilizer; 6) high-
voltage autotransformer; 7) three-phase step-up transformer;
8) rectifier; 9) resistor set (15 kΩ); 10) output stage based on
a GMI-2B tube; 11) discharge-correcting circuits.

A block diagram of the whole apparatus, including the electrical part, is given in Fig. 11.
Electroluminescence was excited by square high-voltage pulses, whose amplitude could be
varied up to 20 kV. Pulses of the following duration could be generated: 1, 1.7, 3 μsec. The
pulse rise time did not exceed 0.1 μsec for the minimum pulse duration and 0.3 μsec for the
maximum pulse duration. The repetition frequency was controlled by an external system of
MGI-2 and GI-3M pulse generators connected in series. This arrangement was used in
combination with a high-voltage pulse generator under single pulse conditions as well as at
low repetition frequencies, beginning from 10 Hz. The high-voltage pulse generator had also
its own triggering unit by means of which one of the following repetition frequencies could be
realized: 300, 500, 1000, 1500, 2500, and 3500 Hz.

The functional circuit of the high-voltage pulse generator is shown in Fig. 12. The
operating principle of the generator was a periodic discharge of a KBG-P-2 capacitor through
an electronic commutation switch. The capacitor was charged from a three-phase rectifier
assembled in accordance with the Larionov scheme using V1-0.1/40 kenotrons. The output
stage (electronic switch) voltage was supplied by bias and screen-grid rectifiers. These
voltages were stabilized by an S-0.75 ferroresonant stabilizer, which operated on the line
supply. The basic pulse-shaping circuit is shown in Fig. 13.

The circuit of the shaping capacitor consisted of a resistor set, three V1-0.1/40
kenotrons connected in parallel in the forward direction, and a milliammeter I_{av}. One of
three plug-in chokes L_1-L_3 was connected in parallel with the kenotrons. An RA-350
discharger was connected in parallel with the milliammeter to protect it from overloads.
The time constant of the charging circuit was thus governed by the capacitance of the shaping
capacitor (0.25 μF) and the charging resistance (15 kΩ). During the charging time a GMI-2B
tube, acting as the electronic switch, was cut off at the first grid by the bias rectifier voltage.
At the moment of arrival of a positive triggering pulse at the first grid the GMI-1B tube
became conducting and provided practically a short-circuit between the shaping capacitor and
the ground. The discharge circuit consisted of a bank of capacitors connected in parallel and
one of the chokes L_1-L_3 (during the discharge the kenotrons were connected in reverse) as
well as the GMI-2B tube itself.

The pulse parameters were governed by the shape of the pulse which unblocked the
electronic switch and by various elements of the discharge circuit. In particular, the pulse

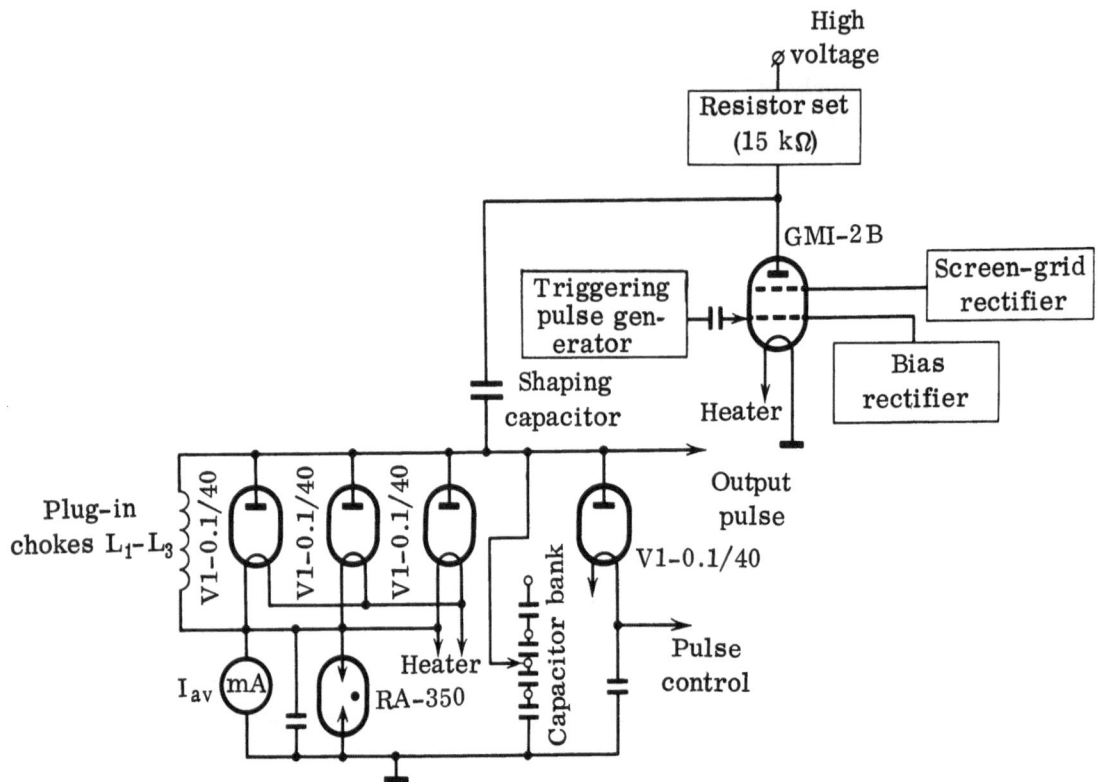

Fig. 13. Basic high-voltage pulse shaping circuit.

rise time could be varied by altering the capacitance of one of the capacitors in the bank. The choke corresponding to the duration of the triggering pulse was selected from the L_1-L_3 set; the pulse rise time could be varied also by altering the voltage applied to the screen grid of the GMI-2B tube. A capacitance-type voltage divider was connected to the output of the discharge correcting circuit to control the pulse shape. The high-voltage arm was the anode—cathode capacitance of a "cold" V1-0.1/40 kenotron.

The voltage across the crystal was measured with a VLI-2 pulse voltmeter and a D62 capacitance divider.

The photomultiplier signal passed through a cathode follower (Fig. 14) based on a 6S15P tube and mounted within the photomultiplier casing. The signal then reached a wide-band amplifier of the USh-2 type. The amplified signal was displayed on the screen of a DÉSO-1 oscillograph. When integrated characteristics were required, the signal was measured with an M-95 microammeter or an ÉPP-09 automatic recorder.

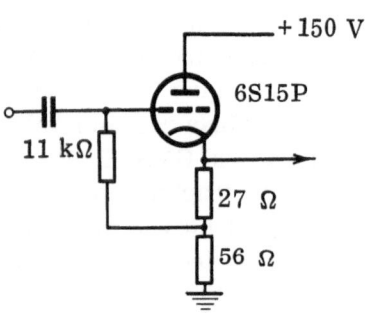

Fig. 14. Cathode follower circuit.

Overheating of the crystal was avoided by using a large off-duty factor ($10^4 - 10^5$). This reduced strongly the signal-to-noise ratio in the recording of the integrated characteristics, such as the luminescence spectrum, the dependence of the electroluminescence brightness on the voltage, etc. Moreover, in a detailed analysis of the spectra we had to use a narrow monochromator slit which also reduced appreciably the signal-to-noise ratio. This ratio was increased by including, between the amplifier and recording instrument (particularly the ÉPP-09 recorder),

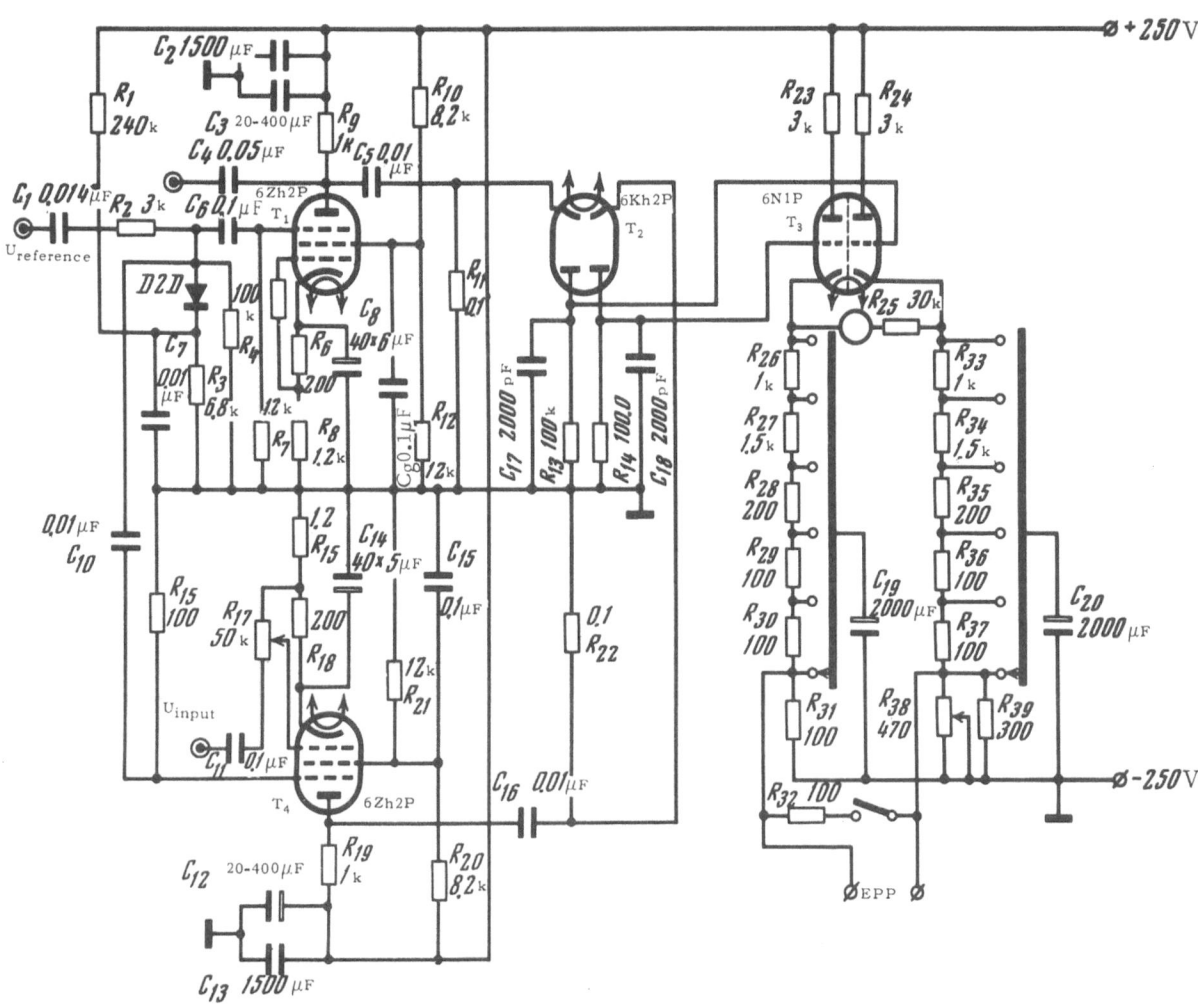

Fig. 15. Synchronous detector circuit.

a synchronous detector specially constructed for this purpose. The circuit of this detector is shown in Fig. 15.

The task of the synchronous detector was to select those parts of the amplifier output signal which contained useful information. This reduced considerably the noise component. The duration of the selected parts of the signal was determined by the duration of a positive reference voltage provided by the MGI-2 pulse generator and applied to the third grid of the tubes T_1 and T_4. The same pulse generator triggered also the high-voltage generator which provided excitation pulses. A signal of negative polarity was applied to the control grid of the tube T_4. In the absence of this signal both stages were controlled by the same reference pulse applied to the third grid and, therefore, both arms of the detector based on a 6Kh2P double diode were subjected to the same conditions. The control grids of a double triode 6N1P (tube T_3) were subjected to the same voltages and the potential difference between the cathode loads (RC circuits) of both parts of the tube T_3 was zero. The appearance of a negative signal at the control grid of the tube T_4 (at the moment when it was triggered by the reference pulse) resulted in blocking of this tube, i.e., in a reduction of the negative potential on the cathode of the tube T_2 and a consequent reduction of the potential on the grid of the left-hand half of the tube T_3. Consequently, the automatic recorder EPP-09 received a signal representing the degree of unbalance of the output stage of the synchronous detector.

The spectra were recorded automatically by means of an attachment to the SPM-1 monochromator (this attachment rotated the monochromator drum at 1 rpm). The spectroscopic calibration of the sensitivity of the apparatus was carried out using a photometric lamp with a known color temperature and the time characteristics of the apparatus were checked by means of a silicon carbide luminescent diode whose response time was 10^{-8} sec.

CHAPTER III

RESULTS OF MEASUREMENTS AND DISCUSSION

§1. Electrical Conductivity

Since the resistivity of zinc sulfide is close to that of dielectrics, special measures must be taken to eliminate surface currents during measurements of the electrical conductivity of this material. In the case of dielectrics this is usually done by means of a guard ring. The same method was applied also in our investigations. The way in which the guard ring was used can be seen from Fig. 9.

Zinc sulfide is known to contain traps of various depths right down to a few tenths of an electron-volt. The photoionization of donors results in electron occupancy of these traps in excess of the thermal value. Since the trap occupancy affects the electrical conductivity, reproducible and true values of the conductivity can be obtained only if a crystal is de-excited by the liberation of electrons from the traps either by heating or by illumination with light of a suitable wavelength. We obtained reproducible results by thermal de-excitation of samples in a light-tight thermostat (see Fig. 12 in the preceding article) by heating them for 15 min at 500°C. This was followed by measurements in the same thermostat.

The samples used in measurements of the electrical conductivity were prepared as described in the preceding chapter. Selected and oriented crystals were cut into samples whose shape and dimensions are shown in Fig. 16. A channel 0.5 mm wide and 0.5 mm deep was cut into a sample to accomodate the guard ring. The actual guard ring was applied in the same way as the contacts supplying the current. In measurements of the temperature dependence of the conductivity the ohmic contacts were coated by silver films.

After de-excitation in the light-tight thermostat the samples had a resistivity $\rho = 10^{14}\ \Omega \cdot$ cm at 440°K [64]. We were unable to measure the resistivity at lower temperatures because this value was outside the limit of sensitivity of our apparatus. The temperature dependence of the conductivity in the 440–770°K range is presented in Fig. 17. It is evident

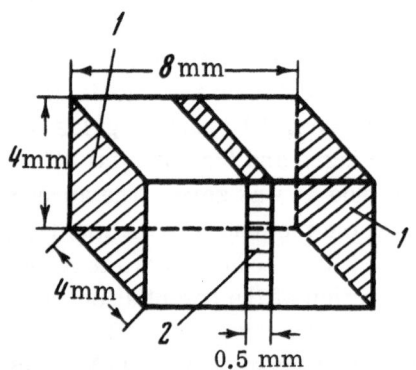

Fig. 16. Shape of the samples used in measurements of the electrical conductivity: 1) electrode; 2) guard ring.

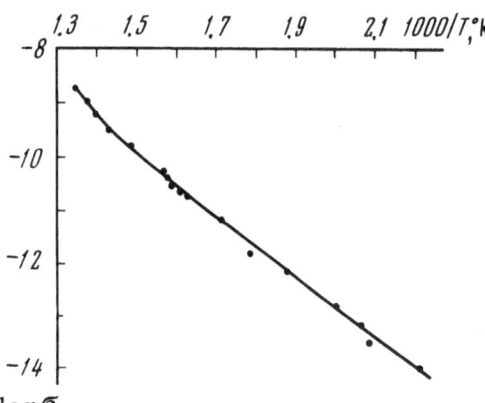

Fig. 17. Temperature dependence of the electrical conductivity.

from this figure that the temperature dependence can be approximated by two straight lines if the coordinates are log σ and T^{-1}. The high-temperature line begins approximately at 700°K and corresponds to an activation energy $\Delta E_1 = 1.6 \pm 0.1$ eV. A high-temperature region, which began at approximately the same temperature and changed to a straight line of the same slope as in our case (within the limits of the experimental error), was observed by Fok [1] for a zinc sulfide powder.

Fok established that the experimental points did not deviate from the straight line right up to 1300°K (this was the highest temperature used in his investigation). For this reason, Fok attributed this range to intrinsic conduction in zinc sulfide. Consequently, our activation energy should be equal to half the thermal width of the forbidden band. The density of free electrons under intrinsic conduction conditions is given by the expression

$$n_i = N_{\text{eff}} \exp\left[-\frac{\Delta E_t}{2kT}\right],$$
(III. 1)

where E_t is the thermal width of the forbidden band, k is the Boltzmann constant, but

$$N_{\text{eff}} = 2\left(\frac{2\pi kT}{h^2}\right)^{3/2}(m_n m_p)^{3/4},$$
(III. 2)

m_n is the effective mass of an electron, and m_p is the effective mass of a hole.

If we assume that the effective masses of an electron and hole are both equal to the free electron mass m, we find that

$$N_{\text{eff}} = 2\left(\frac{2\pi mkT}{h^2}\right)^{3/2}.$$
(III. 3)

The thermal width of the forbidden band is thus $\Delta E_t = 2\Delta E_1 = 3.2 \pm 0.2$ eV, which is in good agreement with the value found by Fok for a zinc sulfide powder [1] . This value agrees also with the thermal width deduced recently from the influence of the temperature on the initial parts of the current−voltage characteristics of p− n junctions in zinc sulfide [65]. This approach yields $\Delta E_t = 3.42 \pm 0.09$ eV. All these values of ΔE_t refer to T = 0°K.

The second line in Fig. 17 corresponds to an activation energy 1.25 ± 0.07 eV and it should be attributed to extrinsic conduction. The density of electrons under extrinsic conduction conditions is described by the following formulas which apply in different cases:

$$n_0 = \frac{N_d - N_a}{2N_a} N_{\text{eff}} \exp\left[-\frac{E_d}{kT}\right],$$
(III. 4)

$$n_0 = N_d^{\frac{1}{2}} N_{\text{eff}} \exp\left[-\frac{E_d}{2kT}\right],$$
(III. 5)

where N_d and N_a are the donor and acceptor concentrations, respectively; E_d is the thermal width of the donor level. These two formulas differ by the coefficient $1/2$ in the argument of the exponential function. Formula (III. 4) applies when $n_0 \ll N_a$ and formula (III. 5) when $n_0 \gg N_a$. In the first case, most of the electrons lost by the donors are captured by the compensation acceptors (the number of vacant donors remains practically constant). In the second case, the number of such electrons is small compared with the number of electrons transferred to the conduction band and, therefore, the number of vacant donors is approximately equal to the number of free electrons. Electrons return to the donors in accordance with the bimolecular reaction law (the probability of donor−electron recombination is proportional to n_0^2). The application of the principle of detailed balance gives rise to a factor $1/2$ in the argument of the exponential function.

The density of free electrons can be estimated using the formula

$$n_0 = \frac{\sigma}{e\mu} , \qquad\qquad\qquad \text{(III. 6)}$$

where μ is the mobility of electrons which is equal to 120 cm$^2 \cdot$ V$^{-1} \cdot$ sec^{-1} [7]. At $T = 440°K$ the conductivity is $\sigma = 10^{-14}$ $\Omega^{-1} \cdot$ cm^{-1} so that $n_0 = 5 \times 10^2$ cm^{-3}. At 770°K the electrical conductivity is about five orders of magnitude higher and, therefore, n_0 reaches $\sim 5 \times 10^7$ cm^{-3}. Even this value is many orders of magnitude lower than the concentration of impurities which can be attained in better semiconducting materials. It follows that the relationship $n_0 \ll N_a$ applies in our case, i.e., formula (III. 4) should be used. The thermal depth of the donors in the investigated crystals of zinc sulfide is equal to the activation energy, i.e., $E_d = 1.25 \pm 0.07$ eV.

Formula (III.4) can be used to estimate the degree of compensation of a crystal $(N_d - N_a)/N_a$. Since $E_d = 1.25$ eV, $N_{eff} = 4 \times 10^{19}$ cm^{-3}, it follows that $(N_d - N_a)/N_a \approx 10^{-2}$. On the other hand, it we assume that the intrinsic conduction region begins at 700°K, we can equate Eqs. (III. 4) and (III. 1) for the extrinsic and intrinsic conduction conditions and again find the value of $(N_d - N_a)/N_a$. Once again, the same value $(N_d - N_a)/N_a \approx 10^{-2}$ is obtained.

We have mentioned earlier that defects in zinc sulfide are generally assumed to be strongly compensated. Our measurements confirm this assumption.

Extrapolation of the low-temperature part of Fig. 17 toward room temperature gives the dark resistivity. The value of this resistivity at room temperature is thus found to be $\sim 10^{20}$ $\Omega \cdot$ cm. However, it is quite likely that zinc sulfide crystals have even shallower donors, whose activation energies may be manifested at temperatures below 440°K. If this is so, the true value of the resistivity at room temperature may differ from the one just given.

In Chap. 1 we have mentioned the strong anisotropy of the electrical conductivity ($\sim 10^6$) and photoconductivity ($\sim 10^4$) of zinc sulfide crystals grown from the vapor phase. It is difficult to explain this anisotropy in the case of perfect ZnS crystals. This is because electronic conduction in ZnS involves zinc ions. Therefore, the anisotropy of the conductivity should be due to a difference in the distances between the zinc ions along different crystallographic directions. Such a difference, however, usually results in an anisotropy of the mobility and of the conductivity which is within one order of magnitude and not six orders, as reported in [10]. Ginter [66] investigated the anisotropy of the mobility in hexagonal cadmium selenide which, like ZnS, belongs to the $A^{II}B^{VI}$ group of compounds. Ginter found that this anisotropy was about 10%.

In view of these contradictions we decided to investigate the anisotropy of the electrical conductivity and photoconductivity of the crystals grown from the melt. Two types of sample were prepared from the same part of a crystal: some were oriented along the C axis and the others at right angles to this axis. The shapes and dimensions of the samples were the same as in Fig. 16. The photoconductivity was excited using a light source of the OI-24 type fitted with a lamp of 100 W power. Such illumination increased the conductivity by eight to nine orders of magnitude. These measurements demonstrated that the electrical conductivity measured parallel to the C axis was always two or three times as large as at right angles to this axis. The photoconductivity along these directions differed by a factor of 1.5.

The conductivity and photoconductivity of different samples cut from neighboring parts of a crystal could differ by a factor of two or three even when the orientation was the same. The scatter of the results of measurements carried out on samples of the same type was less for the samples exhibiting weaker anisotropy of the photoconductivity. This was probably because the photocurrents were large. Therefore, the results obtained by averaging the data obtained for many crystals should be regarded as representing the upper limit of the anisotropy. In view of this, we regarded the photoconductivity anisotropy (a factor of 1.5) as the upper limit of the anisotropy of the photoconductivity and electrical conductivity.

Thus, the anisotropy of the electrical conductivity and photoconductivity was several orders of magnitude smaller (and of opposite sign) than the anisotropy reported in [10]. The anomalously strong anisotropy found by Lempicki et al. [10] was probably due to imperfection of the investigated crystals (these crystals were likely to have oriented dislocations) and possibly due to the particular method used in their measurements.

§2. Visible Electroluminescence

Visible electroluminescence was excited and investigated in melt-grown crystals with Cu concentrations ~ 10^{-6} g-atom/mole. The shape and dimensions of these samples are shown in Fig. 18. The distance between the electrodes was ~ 2 mm. Fairly strong blue photoluminescence was exhibited by these crystals. The room-temperature photoluminescence spectrum is shown in Fig. 19. The excitation wavelength was 360 mμ. The ordinate in Fig. 19 gives the brightness (intensity) I in relative units.

The excitation of electroluminescence in a homogeneous semiconductor with ohmic contacts occurs under conditions favoring the development of electric breakdown (impact and field ionization). However, in the case of large crystals, electric breakdown may be preceded by thermal breakdown which can damage a crystal at voltages insufficient to excite significant electroluminescence. This was why we failed to observe electroluminescence in our crystals by applying the method suitable for the excitation of small ZnS crystals. For this reason, the excitation method was modified as follows. Good heat-removal conditions were established by soldering a crystal to a copper base along one of its ohmic contacts. Overheating of the contact barrier was prevented by heat dissipation via ohmic contacts. However, the most important measure was the use of pulse voltages of about 1 μsec duration and a large (~ 5×10^4) off-duty factor. Sorokina [67] established that the electric breakdown preceded the thermal process (at room temperature) if the pulse duration was of the order of 1 μsec. Sorokina found this to be true even for alkali halide crystals. These measures made it possible to increase considerably the voltages at which thermal breakdown occurred and to observe electroluminescence at voltages corresponding to average fields of ~ 2×10^4 V/cm [68].

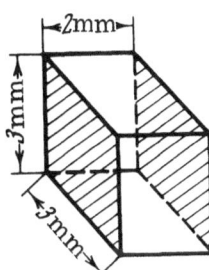

Fig. 18. Shape of the samples used in measurements of the visible electroluminescence. The hatched areas represent electrodes.

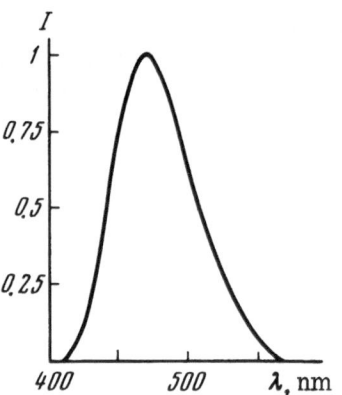

Fig. 19. Photoluminescence spectrum.

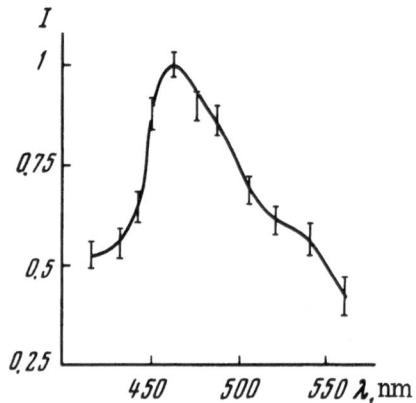

Fig. 20. Visible electroluminescence spectrum.

This luminescence was of the same blue color as the photoluminescence. The electroluminescence spectrum, recorded using a voltage of 6 kV, consisted of a wide band with a maximum at 460 mm (Fig. 20). The examination of electroluminescent crystals under a microscope showed that the electroluminescence was uniform throughout the crystal. The resolving power of the microscope was at least 3 μ. Breakdown in air was prevented by immersing the sample in capacitor oil. The absorption in this oil was allowed for in plotting the electroluminescence spectrum. Control experiments, in which a ZnS sample was replaced by a quartz crystal, showed that the oil itself did not emit electroluminescence.

The average value of the electric field at which electroluminescence was observed was unusually low. This value should be compared with the Chuenkov [42] and Keldysh [43] calculations of the probability of impact ionization which indicated that such ionization should become significant in considerably higher fields $\gtrsim 10^5$ V/cm. The Zener field-ionization mechanism should occur in even stronger fields.

The dependence of the brightness on the voltage is presented in Fig. 21. It is evident from this figure that the dependence is rectilinear if plotted in coordinates log B and U. This means that the dependence can be represented in the form

$$B \propto e^{cU}, \tag{III. 7}$$

where c is a constant independent of the voltage. We note that the nature of this formula does not agree with the formulas predicted for the probability of field ionization [Eq. (I.1)] and impact ionization [Eq. (1.7)]. These formulas are of the type

$$W \propto \exp\left(-\frac{c_1}{E^\gamma}\right),$$

i.e., they have a negative power exponent and the field occurs in the denominator of this exponent. In our dependence the power exponent is positive and the voltage (and therefore the field) occurs in the numerator of the exponent.

There are two field-controlled ionization mechanisms whose probabilities depend on the field in accordance with

$$W \propto \exp(\text{const} \times E^\gamma). \tag{III.8}$$

One is the field ionization, assisted considerably by the thermal vibrations of the lattice, as given by the Keldysh formula of Eq. (I. 3). In this case, $\gamma = 2$. The other mechanism is

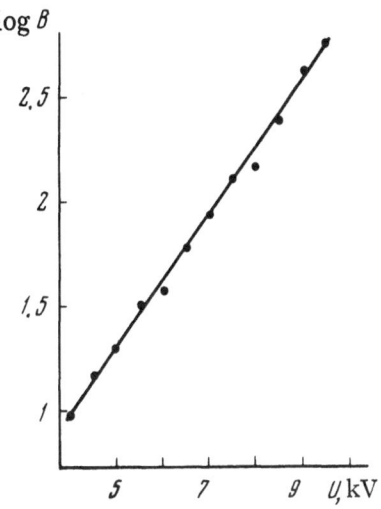

Fig. 21. Dependence of the visible electro-luminescence brightness of the voltage applied to crystals ~ 2 mm thick.

represented by the Frenkel' formula [69]

$$W \propto \exp\left[-\frac{\Delta - 2\sqrt{\frac{e^3 E}{\varepsilon_\infty}}}{kT}\right],$$ (III.9)

where Δ is the activation energy in the absence of a field; e is the electron charge; ε_∞ is the hf permittivity; k is the Boltzmann constant; T is the absolute temperature. This formula rests on the assumption that the ionization potential in the thermal ionization mechanism is reduced considerably by the application of an electric field. In this case, $\gamma = 1/2$. These two processes are intermediate between the thermal and field ionization mechanisms.

There is also a process intermediate between the thermal and impact ionization mechanisms. In strong fields (of the order of the breakdown strength) the thermal vibrations of the lattice impede impact ionization by strong scattering of electrons. However, when a crystal with a considerable ionicity is subjected to a weak field, we may find it necessary to make allowance for the field-induced acceleration of those electrons which have a high mobility after liberation from polaron "clouds" by the thermal vibrations. Unfortunately, this process has not yet been analyzed theoretically.

If we assume that, in our case, $E \propto U$, it follows that $\gamma = 1$ and the results obtained cannot be explained by the Keldysh or Frenkel' mechanisms. Nevertheless, we shall assume that the actual process is due to the simultaneous effects of heat and electric field, i.e., that it is basically impact ionization which is made easier by the lattice vibrations. A combination of several processes is also likely. Moreover, we cannot exclude the possibility of some concentration of the electric field in moving domains, as observed first by Boer [70] in cadmium sulfide. Such domains would have to be relatively "fast" since our voltage pulses were of $\sim 10^{-6}$ sec duration. Fast domains, responsible for low-voltage electro-luminescence in ZnS, have been recently reported by Georgobiani et al. [71].

We have mentioned earlier that unassisted impact ionization requires fields $\geq 10^5$ V/cm. We investigated the electroluminescence in such fields using samples whose working thickness was $\sim 150\ \mu$ (Fig. 22).* In this case, the dependence of the electroluminescence brightness on

*The method used in the preparation of these samples and their shape will be discussed in greater detail in the next section.

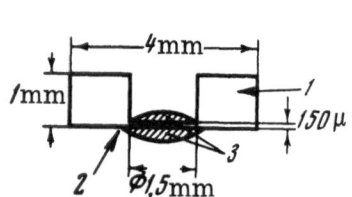

Fig. 22. Shape of the samples used in investigations of the ultraviolet electrolumines-cence. The arrows (1 and 2) indicate the directions of ob-servation. The indium con-tacts are denoted by 3.

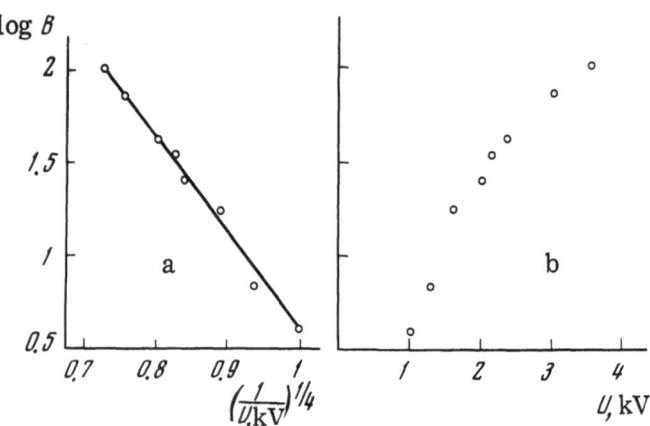

Fig. 23. Dependence of the electroluminescence brightness on the voltage applied to crystals $\sim 150\,\mu$ thick: a) log B plotted against $U^{-1/4}$; b) log B plotted against U.

the voltage was different (Fig. 23). This dependence was rectilinear when plotted in the coor-dinates log B and $U^{-1/4}$. In other words, the dependence could be described by the formula

$$B \propto \exp\left(-\frac{\text{const}}{\sqrt[4]{U}}\right). \qquad \text{(III.10)}$$

This dependence can be explained by impact ionization on the assumption that the field in a crystal is related to the external voltage by

$$E \propto \sqrt[4]{U} \qquad \text{(III. 11)}$$

[this should be compared with Eq. (I. 5)] .

We have mentioned in Chap. I that when the field is concentrated in a Mott — Schottky depletion barrier, the effective field is related to the external voltage by $E \propto \sqrt{U}$. This is a consequence of a uniform distribution of the space charge across the barrier width. Since the contacts of our samples were ohmic, such barriers were not expected to form and we were inclined to assume the presence of moving high-field domains. The charge density distribution in these domains was not known.

A solution of the Poisson equation shows that the dependence (III. 11) may be obtained if the space charge distribution is

$$\rho \propto (L-x)^{-1/3}, \qquad \text{(III.12)}$$

where L is the region over which this charge is distributed. In this case, the distribution of the potential in the space-charge region obeys the formula

$$U \propto (L-x)^{4/3}. \qquad \text{(III.13)}$$

It follows from this discussion that different processes are responsible for the electroluminescence at voltages corresponding to the average fields of 2×10^4 V/cm $\leq E_1 \leq$ 5×10^4 V/cm and $E_2 \geq 10^5$ V/cm. In low fields (E_1) the luminescence centers are ionized by the simultaneous effect of heat and electric field. The impact ionization process is assisted by the thermal energy. In high fields (E_2) impact ionization is not thermally assisted and the effective field is high because it is concentrated in high-field domains.

We investigated the anisotropy of the electroluminescence of zinc sulfide. This was done by preparing samples of two types: some were oriented along the C axis and the others at right angles to this axis. The shape of the samples was the same as in Fig. 18. The electroluminescence emitted by the samples oriented along the C axis was, on the average, 1.5 times stronger than that of the samples oriented at right angles to this axis. However, this difference was within the limits of the scatter of the results obtained for crystals of the same type. The strong anisotropy of the electroluminescence of ZnS single crystals, reported in the published investigations, was evidently due to imperfections in the crystals grown from the vapor phase. These imperfections were also responsible for the strong anisotropy of the electrical conductivity and photoconductivity. This conclusion was drawn from the fact that the crystal structure of ZnS and the associated difference in the ionization potential along different directions could only give rise to a 4% anisotropy even in hexagonal ZnS (Table 1).

§3. Ultraviolet Electroluminescence

The preceding section dealt with the visible electroluminescence. However, the observation and study of the intrinsic luminescence of zinc sulfide are of special interest. The forbidden band width of ZnS corresponds to the ultraviolet part of the spectrum. Since carrier recombination at the luminescence centers may compete successfully with the band–band recombination of electron – hole pairs, the intrinsic luminescence can be observed only in high-quality crystals which contain few impurities and are very close to the stoichiometric composition. Improvements in the crystal growth method, particularly the addition of sulfides to the inert gas, enables us to prepare crystals which had a very weak photoluminescence, were transparent, and had no visible color. On the other hand, much stronger electric fields were required to observe the intrinsic electroluminescence. Such fields were achieved by reducing the thickness of the samples. We prepared samples of special shape suggested by Skanavi [72] for investigating electric breakdown in dielectrics (Fig. 22). Cleaved plates, about 1 mm thick, were cut into squares 4×4 mm area. A recess of 1-1.5 mm diameter was drilled in each such square by ultrasonic machining. The recess had a flat bottom and the thickness of the crystal in the recessed area was 100-150 μ. Indium contacts were applied to both sides of the thin part of the sample by the method described in Chap. II. One of the contacts was soldered to a copper holder. The thick part of the crystal prevented surface breakdown and breakdown in air.

We were able to observe ultraviolet electroluminescence beginning from $\sim 2 \times 10^5$ V/cm [73]. Figure 24 shows the electroluminescence spectrum in the ultraviolet range. It is evident from this figure that the ultraviolet spectrum had two bands with maxima at 335 and 360 nm. The latter band extended right up to 460 nm. The peaks in the visible part of the spectrum were not observed (in contrast to the crystals described in the preceding section).

The brightness of the ultraviolet electroluminescence increased with the voltage somewhat more rapidly than the visible electroluminescence. When the voltage was increased by a factor of 1.7, the brightness of the ultraviolet electroluminescence increased nearly sevenfold, whereas the visible electroluminescence exhibited only a fourfold increase (Fig. 23a). The range of voltages investigated was too narrow to find the functional relationship between the brightness and the voltage. Since the short-wavelength peak ($\lambda_{max} = 335$ nm, $h\nu_{max} = 3.7$ eV) corresponded to the optical width of the forbidden band, we attributed this peak to the band–band recombination of electrons and holes liberated from polarons. The peak ($\lambda_{max} = 360$ nm, $h\nu_{max} = 3.44$ eV) and a considerable part of the long-wavelength band corresponded to the energy interval between the optical and thermal widths of the forbidden band. Consequently, we attributed part of this long-wavelength band to the band–band recombination of electrons and holes but assumed that one or both recombination partners were still in the polaron state.

These results were in good agreement with a recent investigation of the electroluminescence of p – n junctions in zinc sulfide [74]. Forward-biased junctions emitted electroluminescence with a band whose maximum was located at 360 nm (Fig. 22e in the preceding article) but the short-wavelength band was not observed. This was to be expected because the fields in forward-biased junctions were not strong and, therefore, there were no hot electrons in this case. When these junctions were subjected to a reverse bias, the 360 and 335 nm bands were observed.

Ultraviolet luminescence of ZnS can be excited also by a beam of fast electrons [75-78] The reported electroluminescent band positions were somewhat contradictory. This was probably due to the temperature dependence of the forbidden band width or due to differences in the crystal structure of the samples employed. The latter possibility was mentioned by Klein [76] . He found that hexagonal crystals emitted a 320 nm band at 4.2°K, whereas an admixture of the cubic phase shifted this band to 330 nm. Bogdankevich et al. [75] investigated melt-grown single crystals and found that bombardment with fast electrons at room temperature excited a band with a maximum at 345 nm. At 80°K Bogdankevich et al. observed bands at 330, 335, and 345 nm when the current density was about 6 A/cm². Arapova et al. [77] found that electron-beam currents at 10^{-6} A/cm² density were sufficient to excite 335, 338, and 342 nm bands at T = 80°K in powder samples. Hurwitz [78] found that electron bombardment of ZnS generated stimulated radiation of 329 nm wavelength when the current density was only 0.1 A/cm².

Comparisons of electroluminescence with cathodoluminescence must be treated with caution, particularly when electroluminescence is investigated during the application of an electric field pulse. This is because a strong electric field can alter the electron energy spectrum by the Stark effect and it can alter considerably the energy distribution of free electrons and holes.

Samples prepared from one of the crystals at our disposal (this crystal was grown by L. A. Sysoev and S. A. Fridman) exhibited an additional (very weak) luminescence peak at about 315 nm. This peak was observed along the direction represented by arrow 2 in Fig. 22 but not along the direction represented by arrow 1. This was evidently due to absorption in the fundamental region. The energy of the quanta corresponding to this peak exceeded considerably the optical width of the forbidden band and, therefore, it was natural to attribute the peak to the recombination of "hot" electrons and holes. The position of this peak was evidently associated with some singularity in the allowed band structure.

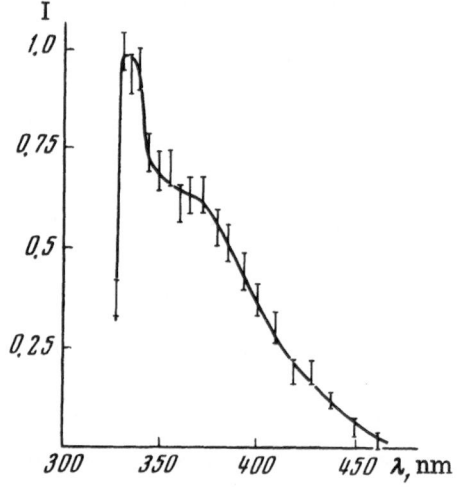

Fig. 24. Ultraviolet electroluminescence
spectrum.

CONCLUSIONS

We were able to observe the electroluminescence of homogeneous zinc sulfide samples which were free of internal and contact barriers. The luminescence was localized in separate lines in small crystals grown from the vapor phase because these crystals had large defects, particularly dislocations and internal heterojunctions.

Crystals which exhibited strong blue photoluminescence emitted also blue electroluminescence at voltages corresponding to an average field of $\sim 2 \times 10^4$ V/cm in a crystal. In such fields the dependence of the brightness on the voltage was approximately given by the formula

$$B \propto \exp\left(\text{const} \times U\right),$$

whereas in fields $\sim 10^5$ V/cm this dependence was

$$B \propto \exp\left(\frac{\text{const}}{\sqrt[4]{U}}\right).$$

The transition from the former to the latter dependence was due to the transition from the impact ionization of the luminescence centers by the simultaneous effect of heat and electric field to the "pure" impact ionization process.

We observed also ultraviolet electroluminescence of zinc sulfide. This electroluminescence was found in crystals whose photoluminescence was very weak. The ultraviolet luminescence spectra had peaks at 360, 335, and 315 nm. All these peaks were attributed to the band — band recombination of electron—hole pairs .

An investigation of the temperature dependence of the electrical conductivity enabled us to determine the thermal width of the forbidden band which was $\Delta E_t = 3.2 \pm 0.2$ eV. This value agreed, within the limits of the experimental error, with the forbidden band width of zinc sulfide found for powders [1] and crystals with p — n junctions [65].

We also determined the depth of the donor levels which was $E_d = 1.25 \pm 0.07$ eV. The electrical conductivity data confirmed the hypothesis that the degree of compensation of ZnS was very high. The value of $(N_d - N_a)/N_a$, determined by two different methods, was approximately 10^{-2}.

The absolute value of the resistivity could be measured only at relatively high temperatures of at least 440°K. At this temperature the resistivity was $\rho \approx 10^{14}\ \Omega \cdot$ cm. We found that the anisotropy of the electrical conductivity, photoconductivity, and electroluminescence of homogeneous ZnS crystals was relatively weak. The anomalously strong anisotropy of ZnS, reported by other workers, was evidently due to the presence of imperfections (oriented dislocations) in their crystals.

The authors are grateful to L. A. Sysoev for kindly supplying the crystals, and to M. I. Elinson, Ya. A. Oksman, M. V. Fok, and V. A. Chuenkov for discussing the results and for their valuable comments.

LITERATURE CITED

1. M. V. Fok, Fiz. Tverd. Tela, 5:1489 (1963).
2. M. Cardona and G. Harbeke, Phys. Rev., 137:A1467 (1965).
3. G. S. Zhdanov, Crystals Physics, Oliver and Boyd, Edinburgh (1965).

4. B. D. Saksena, Phys. Rev., 81:1012 (1951).

5. P. F. Browne, J. Electron., 2:154 (1956).

6. B. V. Nekrasov, Course of General Chemistry /in Russian/, Goskhimizdat, Moscow (1952).

7. P. Zalm, Philips Res. Rep., 11:353 (1956).

8. R. H. Bube, Photoconductivity of Solids, Wiley, New York (1960).

9. A. N. Georgobiani and V. I. Steblin, Opt. Spektrosk., 22:167 (1967).

10. A. Lempicki, D. R. Frankl, and V. A. Brophy, Phys. Rev., 107:1238 (1957).

11. W. W. Piper, Phys. Rev., 92:23 (1953).

12. E. N. Arkad'eva, Fiz. Tverd. Tela 6:1034 (1964).

13. M. N. Alentsev and E. I. Panasyuk, Opt. Spektrosk., 5:207 (1958).

14. E. F. Gross and L. G. Suslina, Opt. Spektrosk., 6:115 (1959); Izv. Akad. Nauk SSSR, Ser. Fiz., 25:532 (1961).

15. A. N. Georgobiani and Kh. Fridrikh, Fiz. Tverd. Tela (in press); A. N. Georgobiani and Kh. Fridrikh, Abstracts of Papers presented at Second all-Union Conf. on $A^{II}B^{VI}$ Compounds, Uzhgorod, 1969 [in Russian], p. 61.

16. M. L. Cohen and T. K. Bergstresser, Phys. Rev., 141:789 (1966).

17. J. Birman, Phys. Rev., 109:810 (1958).

18. K. V. Shalimova, Doctoral Thesis [in Russian], Physics Institute, Academy of Sciences of the USSR, Moscow (1952).

19. K. V. Shalimova and N. K. Morozova, Izv. Vyssh. Ucheb. Zaved., Fizika, No. 2, 98, 104 (1964).

20. É. Ya. Arapova, Opt. Spektrosk., 13:416 (1962).

21. W. W. Piper and F. E. Williams, Solid State Phys., 6:95 (1958).

22. K.-S. K. Rebane, Luminescence [in Russian], Vol. 2, Tartu (1966).

23. J. S. Prener and F. E. Williams, Phys. Rev., 101:1427 (1956).

24. J. S. Prener and F. E. Williams, J. Phys. Radium, 17:667 (1956).

25. J. S. Prener and F. E. Williams, J. Chem. Phys., 25:361 (1956).

26. H. Samelson and A. Lempicki, Phys. Rev., 125:901 (1962).

27. J. S. Prener and D. J. Weil, J. Electrochem. Soc., 106:409 (1959).

28. I. Uchida, J. Phys. Soc. Jap., 19:670 (1964).

29. S. Larach and R. E. Shrader, RCA Rev., 20:532 (1959).

30. H. C. Froelich, J. Electrochem. Soc., 100:280 (1953).

31. B. N. Favorin, G. S. Kozina, and L. K. Tikhonova, Opt. Spektrosk., 7:703 (1959).

32. G. S. Kozina and L. P. Poskacheeva, Opt. Spectrosk., 8:214 (1960).

33. J. N. Bowtell and H. C. Bate, Proc. IRE, 44:697 (1956).

34. D. W. G. Ballentyne, J. Phys. Chem. Solids, 10:242 (1959).

35. C. Zener, Proc. Roy. Soc., London, A145:523 (1934).

36. L. V. Keldysh, Zh. Eksp. Teor. Fiz., 34:962 (1958).

37. N. A. Vlasenko, Opt. Spektrosk., 7:511 (1959).

38. F. Möglich and R. Rompe, Z. Phys., 119:472 (1942).

39. C. Z. van Doorn, Physica, 20:1155 (1954).

40. C. K. Coogan, Proc. Phys. Soc., London, B70:845 (1957).

41. H. B. Callen, Phys. Rev., 76:1394 (1949).

42. V. A. Chuenkov, Fiz. Tverd. Tela Sbornik (Suppl.), Vol. 2, pp. 200, 209 (1959).

43. L. V. Keldysh, Zh. Eksp. Teor. Fiz., 37:713 (1959).

44. L. V. Keldysh, Zh. Eksp. Teor. Fiz., 48:1692 (1965).

45. V. A. Chuenkov, Fiz. Tverd. Tela, 9:48 (1967).

46. A. Levialdi and E. Guercigh, C. R. Acad. Sci., 257:852 (1963).

47. Yu. P. Chukova, Tr. Fiz. Inst. Akad Nauk SSSR, 37:149 (1966) [Electrical and Optical Properties of Semiconductors, Consultants Bureau, New York (1968) p. 127].

48. G. Diemer, Philips Res. Rep., 10:194 (1955).

49. E. E. Loebner and H. Freund, Phys. Rev., 98:1545 (1955).
50. W. Lehmann, J. Electrochem. Soc., 107:657 (1960).
51. W. Lehmann, J. Electrochem. Soc., 104:45 (1957).
52. G. Destriau, Symp. Brooklyn Polytech. Inst., Brooklyn (Sept. 1955).
53. A. N. Georgobiani and M. V. Fok, Opt. Spektrosk., 10:188 (1961).
54. P. Zalm, G. Diemer, and H. A. Klasens, Philips Res. Rep., 10:205 (1955).
55. A. N. Georgobiani and V. I. Steblin, Fiz. Tekh. Poluprov., 1:931 (1967).
56. V. E. Oranovskii, E. I. Panasyuk, and B. T. Fedyushin, Inzh.-Fiz. Zh., 2(1):39 (1959).
57. N. A. Vlasenko and Yu. A. Popkov, Opt. Spektrosk., 8:81 (1960).
58. N. G. Basov, B. D. Osipov, and A. N. Khvoshchev, Zh. Eksp. Teor. Fiz., 40: 1882 (1961).
59. V. A. Sokolov, and R. V. Gurova, Proc. Intern. Conf. on Luminescence, Budapest, 1966,
    Vol. 2, publ. by Akadémiai Kiadó, Budapest (1968), p. 1981.
60. S. S. Yee, Appl. Phys. Lett., 9:10 (1966).
61. H. Hartmann, Phys. Status Solidi, 2:585 (1962).
62. H. Gobrecht, H. Nelkowski, and P. Albrecht, Z. Naturforsch., 16a:857 (1961).
63. I. Cooke, J. Chem. Phys., 38:291 (1963).
64. Yu. V. Bochkov, A. N. Georgobiani, and G. S. Chilaya, Fiz. Tverd. Tela, 8:1273 (1966).
65. A. N. Georgobiani and V. I. Steblin, Radiotekh. Elektron., 13:1068 (1968).
66. I. Ginter, Dissertation for Candidate's Degree, Warsaw (1966).
67. L. A. Sorokina, Dissertation for Candidate's Degree [in Russian], Physics Institute,
    Academy of Sciences of the USSR, Moscow (1963).
68. Yu. V. Bochkov, A. N. Georgobiani, I. I. Kisil', L. A. Sysoev, and G. S. Chilaya, Izv.
    Akad. Nauk SSSR, Ser. Fiz., 30:629 (1966).
69. Ya. I. Frenkel', Zh. Eksp. Teor. Fiz., 8:1292 (1938).
70. K. V. Ber (K. W. Boer), Izv. Akad. Nauk SSSR, Ser. Fiz., 24:43 (1960).
71. A. N. Georgobiani, A. V. Lavrov, P. A. Todua, and V. A. Chikhacheva, Abstracts of
    Papers presented at Third All-Union Conf. on Electroluminescence, Tartu, 1969
    [in Russian], p. 94.
72. G. I. Skanavi, Physics of Dielectrics [in Russian], 2 Vols., Gosizdat, Moscow-Leningrad
    (1958).
73. Yu. V. Bochkov, A. N. Georgobiani, A. S. Gershun, L. A. Sysoev, and G. S. Chilaya, Opt.
    Spektrosk., 20:183 (1966).
74. A. N. Georgobiani and V. I. Steblin, Fiz. Tekh. Poluprov., 1:934, 956 (1967).
75. O. V. Bogdankevich, M. M. Zverev, A. I. Pechenov, and L. A. Sysoev, Fiz. Tverd. Tela,
    8:2547 (1966).
76. W. Klein, J. Phys. Chem. Solids, 26:1517 (1965).
77. É. Ya. Arapova, Yu. V. Voronov, V. L. Levshin, V. A. Chikhacheva, and V. V. Shchaenko,
    Izd. Akad. Nauk SSSR, Ser. Fiz., 30:1490 (1966).
78. C. E. Hurwitz, Appl. Phys. Lett., 9:116 (1966).

TEMPERATURE DEPENDENCE OF
THE BRIGHTNESS OF THE RADIATION EMITTED
BY ELECTROLUMINESCENT CAPACITORS

Yu. P. Chukova

A simultaneous analysis was made of the temperature dependences of the brightness of the radiation emitted by an electroluminescent capacitor and of the current rectified by this capacitor. The influence of the frequency of the excitation field on these characteristics was also determined. This made it possible to draw some conclusions about the mechanism which governed the electroluminescence brightness. It was concluded that the physical meaning of the parameters deduced from the temperature dependence of the brightness should be approached with great caution.

The temperature dependence of the electroluminescence brightness is an important characteristic of the Destriau effect. This characteristic is important not so much because of the applications of this effect (the luminescence brightness varies little in the range of temperatures in which electroluminescent panels are operated) but because it can yield interesting information on the mechanism of electroluminescence. This interest is not surprising because there are still many obscure aspects of this mechanism, for example, the relationship between impact and tunnel ionization in the excitation of an electroluminescent phosphor. Some workers [1] attribute the excitation of electroluminescent phosphors simply to impact ionization in a p−n junction subjected to a reverse bias, whereas others [2] associated the average electroluminescence brightness with the tunnel effect. Several specific mechanisms of electroluminescence have been suggested and the results of calculations based on appropriate models have been compared with the experimental data. Although different authors have used different models and have obtained different temperature dependences of the brightness, each has concluded, nevertheless, that his calculations are in agreement with experiment. This has happened because there is a great variety of experimentally obtained temperature dependences of the average brightness. Such experimental data have been combined with various models in the determinations of some of the parameters of electroluminescent zinc sulfide, such as the depth of traps [3, 4], the frequency factor [5], the phonon energy [6], etc. The values of the parameters obtained in this way are not always in agreement and this raises doubts about the correctness of the method used to determine these parameters. Moreover, such contradictions suggest that there may be other ways of explaining the mechanism of electroluminescence.

With these points in mind, the present author investigated many characteristics of an electroluminescent capacitor, including the dependences of the absorbed power, brightness waves, and rectified current [7] on the frequency, voltage, and temperature. The investigations of the current rectified by an electroluminescent capacitor, excited by an alternating electric field, were particularly successful.

Fok and Chukova [8] demonstrated that phonon-assisted tunnel leakage of electrons occurred at $77°K < T < 170°K$ in heterojunctions made of electroluminescent zinc sulfide. In this range of temperatures the temperature dependence of the current $\bar{\mathcal{J}}$, rectified by an electroluminescent capacitor was in good agreement with the probability of single-phonon tunnel creation of pairs n_0 [9]:

$$n_0 = F \left\{ N_k + (1 + N_k) \exp \left[-\frac{4\sqrt{2m^* \Delta}}{e\hbar E} \hbar\omega \right] \exp \left[-\frac{4\sqrt{2m^*}}{3e\hbar E} (\Delta - \hbar\omega)^{1/2} \right] \right\}, \tag{1}$$

where F is a weak function of the temperature and field; m^* is the effective mass of carriers; $\hbar\omega$ is the energy of a lattice phonon; E is the field intensity; Δ is the height of the heterojunction barrier; N_k is the Planck distribution given by

$$N_k = \left[\exp \left(\frac{\hbar\omega}{kT} \right) - 1 \right]^{-1}. \tag{2}$$

An investigation of the characteristics of the rectified current at liquid nitrogen temperatures made it possible to determine the following parameters of the barrier: the barrier height 0.34 eV, the field intensity in the barrier $E \approx 10^6$ V/cm, the concentration of ionized impurities $N \approx 10^{17}$ cm^{-3}, the barrier width $L \approx 10^{-5}$ cm, and the energy (0.034 eV) of phonons participating in the tunnel transitions.

The purpose of the present investigation was to determine whether the conclusions drawn from the study of the rectified current [8] apply also to the temperature dependence of the brightness. An analysis was also made of whether it would be possible to determine some parameters from the temperature dependences.

Method

Temperature dependences were obtained for two characteristics of an electroluminescent capacitor: the average brightness B and the current $\bar{\mathcal{J}}$ rectified by the capacitor excited with an alternating sinusoidal voltage. The measurements were carried out in a thermostat described earlier [10]. The apparatus made it possible to determine simultaneously B and $\bar{\mathcal{J}}$.

A VÉI-2 photomultiplier with an antimony—cesium photocathode was used as the radiation detector. The green band in the electroluminescence spectrum was isolated by an OS-11 filter and the blue band was isolated by means of FS-7 and SS-4 filters.

The samples were prepared in two ways, depending on the binder used. If the binder was an ÉP-96 epoxy resin or a mixture of polystyrene with nitrocellulose, the samples were prepared by spraying a suspension of the phosphor in the binder on the surface of a glass substrate. A different method was used only if Terylene was employed as the binder. In this case, a Terylene film, 10 μ thick, was placed on cold polished glass coated with a conducting layer of SnO$_2$. The glass and Terylene were placed in an oven in which the temperature was raised gradually to 250°C. This melted the Terylene film which wetted uniformly the glass surface. The glass was cooled slowly and a suspension of the phosphor in alcohol was sprayed onto the Terylene film. When the alcohol had evaporated, a second piece of Terylene film was deposited on the phosphor layer and the sample was heated as before. A third layer was then superimposed on this sandwich and the procedure repeated. An aluminum electrode was evaporated on top of the sample.

The thickness of the investigated samples ranged from 50 to 100 μ, depending on the preparation method. The average area of the sample was 7×7 cm.

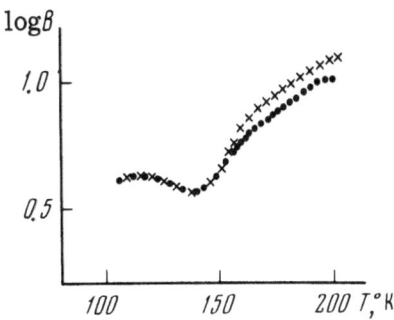

Fig. 1. Temperature dependence of the brightness B for
samples PN-8 (points) and PN-
10 (crosses); ν = 50 Hz, U =
100 V.

In most cases, we used ZnS:Cu phosphors of the ÉL-
510 and ÉL-510^{m} grades. These phosphors emitted green
luminiscence. A few of the samples were prepared from
zinc sulfide of the ÉL-460 grade which emitted blue luminescence. The samples prepared from Terylene were designated
by the letter T, those prepared from a mixture of polystyrene
with nitrocellulose were designed by P and PN, and those
prepared using the ÉP-96 resin were denoted by R.

Experimental Results

The published temperature dependences of the electroluminescence brightness differ considerably even in the case
of phosphors, i.e., copper-activated zinc sulfide. Some authors have observed a monotonic rise of the brightness [6, 11]
between liquid nitrogen and room temperatures, whereas
others have found brightness maxima and minima in the same temperature range [5, 12]. In
spite of these very large differences between the experimentally determined temperature dependences of the brightness, no analysis of the causes of these differences has yet been made
and even comparisons of two types of temperature dependence are lacking. Therefore, we must
consider first the reproducibility of the results obtained for a phosphor of the same grade.
Figure 1 shows the temperature dependences for two samples prepared from the ÉL-510 phosphor suspended in a mixture of polystyrene and nitrocellulose. The reproducibility of the results for samples of this type was satisfactory. Since these samples were basically heterogeneous systems, it seemed desirable to determine the influence of the binder on the temperature
dependence B = B(T). It is evident from Fig. 2a that the same dependences B = B(T) were
obtained for samples prepared using the epoxy resin or a mixture of polystyrene and nitrocellulose. The minima and maxima for samples R-308 and P-212 were observed at the same
temperatures and only the absolute value of the brightness was different. Figure 2b shows the
results obtained for samples made up from Terylene and a mixture of polystyrene and nitrocellulose. The positions of the extrema were still the same but the shapes of the curves were

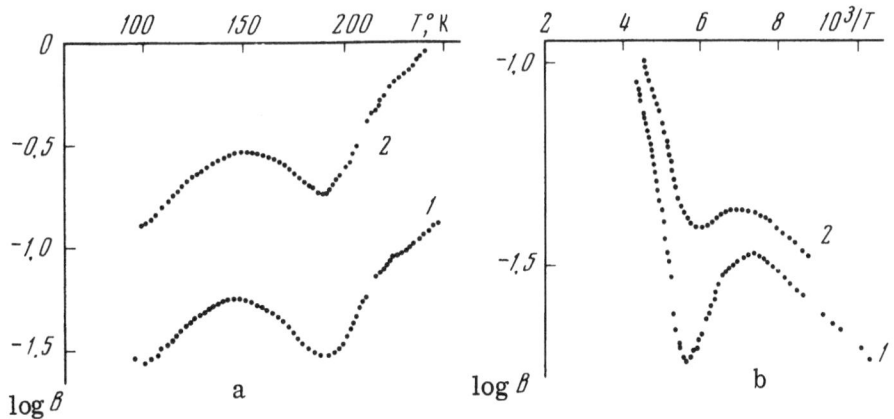

Fig. 2. Temperature dependence of the brightness for samples made
using different binders: a) sample P-212 with polystyrene + nitrocellulose binder (1) and sample R-308 with ÉP-96 resin binder (2), U =
200 V, ν = 15 kHz; b) sample P-212 (1) and sample T-8 (2), U = 200 V,
ν = 5 kHz.

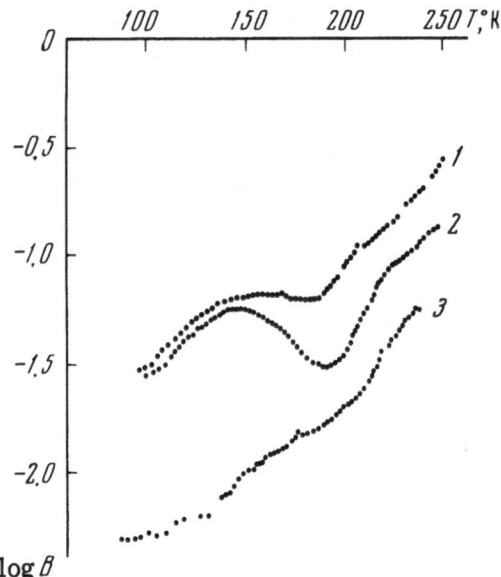

Fig. 3. Temperature dependence of the brightness for samples PN-15 (1), P-212 (2), and PN-14 (3).

different because of the different slopes in the middle region. The slopes of the outer ends of the dependences B = B(T) as well as the positions of the extrema were found to be insensitive to the nature of the binder.

We concluded that, although the type of binder affected the absolute value of the brightness, it did not alter the basic nature of the temperature dependence of the brightness. The results presented in Fig. 2b indicated that the parameters of an electroluminescent material could be deduced from the temperature dependence of the brightness only if independent experiments demonstrated that the dielectric properties of the binder did not vary with temperature. The temperature dependence of the brightness was determined for a large number of samples and it was found that some of them exhibited a monotonic rise of the brightness with increasing temperature whereas others exhibited maxima and minima (Fig. 3). The curves differed most between 150 and 200°K. In this temperature range the brightness of the electroluminescence emitted by sample PN-15 was almost independent of the temperature, whereas sample PN-14 exhibited a rise of the brightness and sample P-212 exhibited a strong fall.

The most surprising observation was that these very different characteristics were obtained for samples prepared by the same method using the same binder and the same electroluminescent phosphor of the ÉL-510 grade (VTU No. 3-384-60) prepared at the "Red Chemist" factory. The only difference between these samples was that different batches of the phosphor were used. Samples PN-14 and PN-15 were prepared from the ÉL-510 phosphor of the batch manufactured on April 9, 1961, whereas sample P-212 was prepared from the ÉL-510 phosphor manufactured on June 6, 1962. Both batches were manufactured in accordance with the same VTU No. 3-384-60 specification. The phosphors prepared from different batches manufactured after 1962 (ÉL-510 and ÉL-510^m grades) exhibited minima and maxima. Since the phosphors prepared in accordance with the same specification had quite different temperature dependences B = B(T), we concluded that the temperature dependence of the electroluminescence brightness was a very sensitive characteristic of the process used in the manufacture of a phosphor and, therefore, of detailed aspects of the electroluminescence mechanism.

Since an electroluminescent capacitor is a complex heterogeneous system, the average characteristics of such a capacitor (for example, the brightness of its electroluminescence) are values which are averages over many parameters (time, coordinates, particle size, etc.) and the resultant curves represent the superposition of many elementary curves. Variations in the temperature dependence of the brightness may be due to changes in the granulometric composition of ZnS powders. This hypothesis seems to be supported by the fact that phosphors of the ÉL-510^m grade, which consist only of fine particles, always exhibit non-monotonic temperature dependences of the brightness.

It seemed of interest to determine the influence of external factors on the temperature dependence of the brightness. The influence of the frequency of the excitation field was investigated further. The results obtained are presented in Fig. 4 and 5. Figure 4 shows the temperature dependence of the brightness for sample PN-14 excited by an external voltage

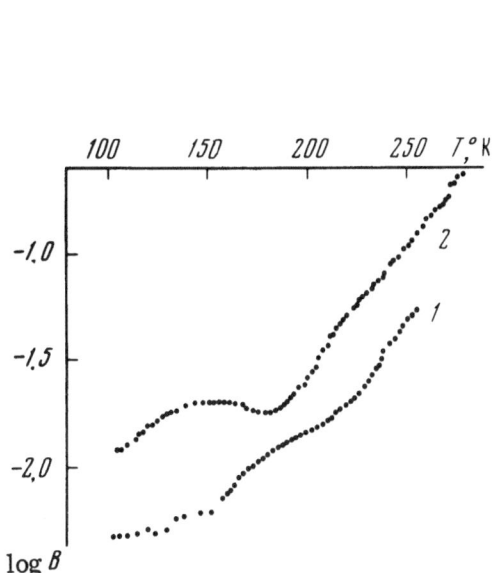

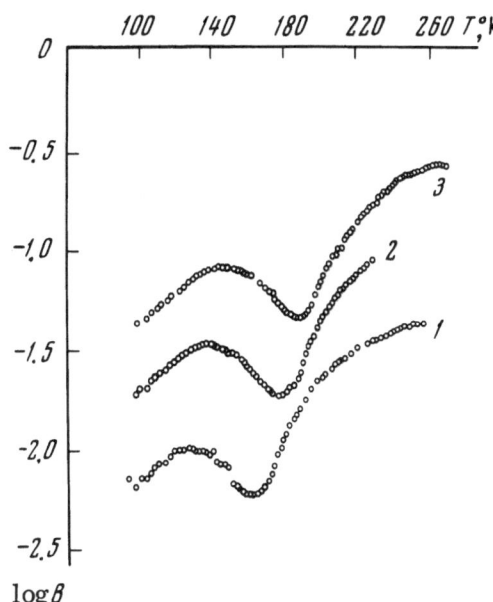

Fig. 4. Temperature dependence of the brightness for sample PN-14 excited by a voltage U = 100 V of 200 Hz (curve 1) or 15 kHz (curve 2) frequency.

Fig. 5. Temperature dependence of the brightness for sample P-212 excited with U = 200 V, ν = 1 (1), 5 (2), and 12 kHz (3).

U = 100 V of frequency ν = 200 Hz or 15 kHz. We can see that when the frequency is increased the temperature dependence becomes nonmonotonic. Figure 5 shows the results obtained for sample P-212 at a voltage U = 200 V and three frequencies. The nature of the curve is independent of the frequency but the whole temperature dependence shifts in the direction of higher temperatures with increasing frequency.

These results indicate that the elementary temperature dependences of the brightness are those shown in Fig. 5. The superposition of such curves can give monotonic dependences of the type shown in Fig. 4. Hence, it follows that any theory which ignores the averaging of the characteristics should be compared with the B = B(T) curves with maxima and minima and not with the monotonic curves as done in [11]. Therefore, we shall restrict our analysis to the nonmonotonic dependences B = B(T).

It would be very attractive to explain the temperature dependence of the brightness by the mechanism of the single-phonon tunnel effect on the basis of the results reported in [8] for the rectified current. The similarity of the electroluminescence brightness and of the rectified current of an electroluminescent capacitor has been mentioned in [7]. These two characteristics have the same dependences on the external voltage and on the frequency of that voltage. However, a more detailed comparison of the results reported in [10, 13] indicated also some differences. An analysis of these differences established that the effective field in current rectification differed from the effective field in electroluminescence. The former was smaller than the latter and it depended on the frequency, whereas the latter was constant in a wide range of frequencies (at a fixed value of the voltage). Therefore, it seemed interesting to compare the temperature dependences of the brightness and of the rectified current.

We can see from Fig. 6a that the brightness and the current for sample P-248[a] (ÉL-510, batch manufactured on April 9, 1961) increased with increasing temperature. Of special interest were the temperature dependences obtained for sample P-26 (ÉL-510, batch manufactured on June 6, 1962) which are presented in Fig. 6b.

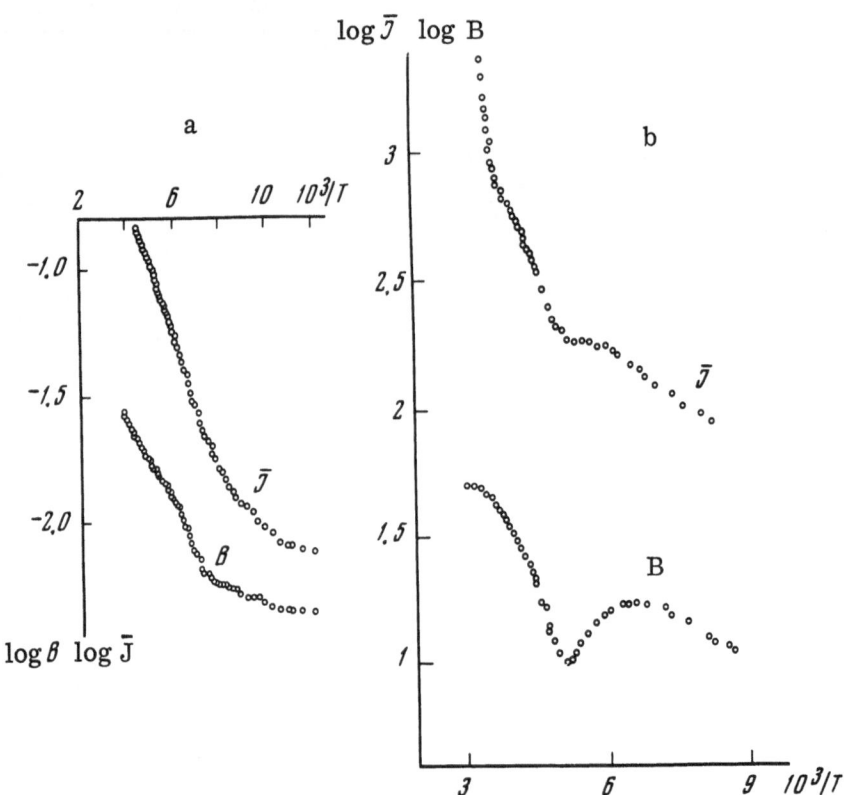

Fig. 6. Temperature dependences of the rectified current $\bar{J}$ and
of the luminescence brightness B: a) sample P-248[a] , ν = 25 Hz,
U = 100 V; b) sample P-26, ν = 12 kHz, U = 100 V.

At temperatures T < 200°K the dependences of the brightness and rectified current for
sample P-26 did not agree: $\bar{J}$ varied monotonically with the temperature whereas the brightness B passed through a maximum. At temperatures T > 200°K (the actual boundary between
these two regions depended strongly on the frequency of the excitation field) the brightness and
the current increased proportionally to one another up to ≈ 270°K. The temperature dependences, plotted in the coordinates

$$\log \bar{J} = f\left(\frac{1}{T}\right) \text{ and } \log B = f\left(\frac{1}{T}\right) ,$$

were practically linear at high temperatures. The slopes of the temperature dependences of the
brightness and current in the 200°K < T < 270°K range were approximately equal. The results
obtained are presented in Table 1 for U = 100 V. At higher temperatures the brightness increased slowly with the temperature, whereas the current $\bar{J}$ increased rapidly. This relationship between the brightness B and the rectified current $\bar{J}$ was typical of the ÉL-510 and
ÉL-510[m] phosphors and was obtained for a large number
of samples.

TABLE 1

Sample	Frequency	Slope of B, °K	Slope of $\bar{J}$, °K
P-26	500 Hz	340	360
P-237	12 kHz	400	380
P-232	12 kHz	390	370

Our experiments showed that the difference in
the behavior of the current and brightness near room
temperature was due to an increase in the loss-angle
tangent of the binder (polystyrene + nitrocellulose).
The brightness was practically independent of the loss-
angle tangent, whereas the current was, to a consider-

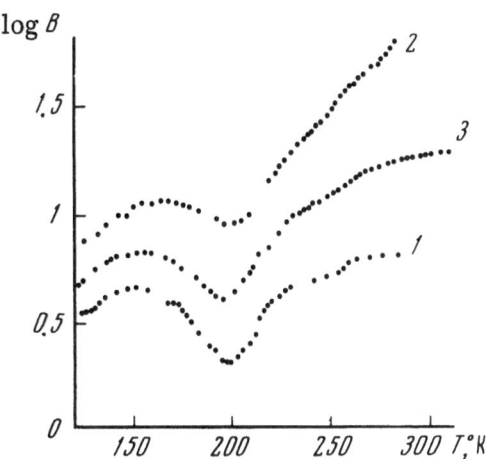

Fig. 7. Temperature dependence of the brightness for sample P-26 (U = 100 V, ν = 12 kHz): 1) blue band; 2) green band; 3) integrated brightness.

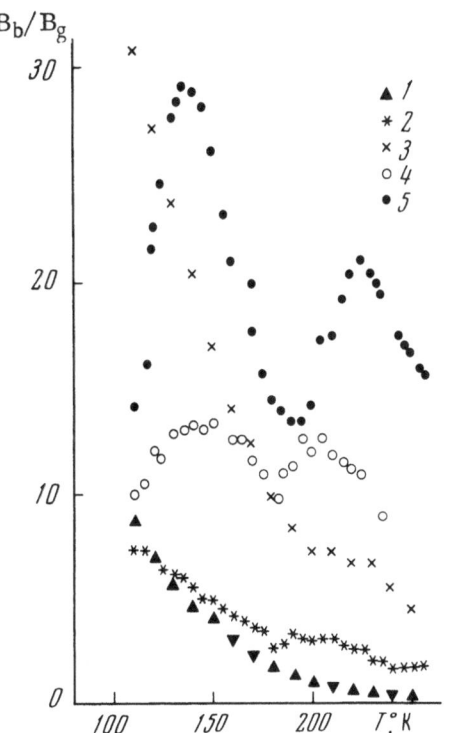

Fig. 8. Temperature dependence of the ratio of the blue band to the brightness of the green band: 1) P-22 [phosphor No. 142(3)], 50 Hz; 2) P-13 (phosphor ÉL-510, batch manufactured on April 9, 1961); 3) P-13, 20 kHz; 4) P-29 (phosphor ÉL-510, batch manufactured on June 6, 1962), 5 kHz; 5) P-29, 20 kHz.

able extent, determined by the conductance of the sample in accordance with the formula

$$\overline{\jmath} = \frac{\varepsilon}{R_0 + R_\infty + R + R_b} , \qquad (3)$$

where ε is the constant emf generated at the barrier in the capacitor; R_0 is the barrier resistance; R_∞ is the bulk resistance of ZnS:Cu; R is the electrode resistance; R_b is the resistance of the binder [7].

The luminescence spectra of the samples prepared from the ÉL-510 and ÉL-510^m phosphors were mixtures of green and blue bands. The color of the electroluminescence depended on the temperature: at liquid nitrogen temperature it was blue and at room temperature it was green. The results given in the preceding paragraph were obtained for the integrated brightness (brightness of all the electroluminescence irrespective of its color) but additional measurements were made to determine the temperature dependences of the brightness of each of the two bands. The temperature dependences of the brightness of the blue and green bands and of the total brightness are presented in Fig. 7 for sample P-26. The brightness is plotted in Fig. 7 in relative units and the curves are placed in an arbitrary manner along the ordinate. The extrema of all three curves were located in the same narrow range of temperatures but the minimum and maximum of the green band were shifted somewhat in the direction of higher temperatures.

The temperature dependences of the brightness of the blue (B_b) and green (B_g) bands were compared by plotting the temperature dependences of the ratio B_b/B_g for various frequencies of the field (Fig. 8). The values along the ordinate (representing the ratio B_b/B_g) are only relative because the brightness of the two bands was measured in relative units. These results indicated that the minimum and maximum in the temperature dependence of the average brightness B were not due to a redistribution of recombining carriers between two luminescence bands. However, at low frequencies (Fig. 8), the migration of holes from the blue to the green centers was observed in sample P-22 when the temperature was increased. Therefore, an additional check of the hypothesis that a redistribution of carriers did not occur was made using phosphor No. 142(3) (this phosphor was prepared

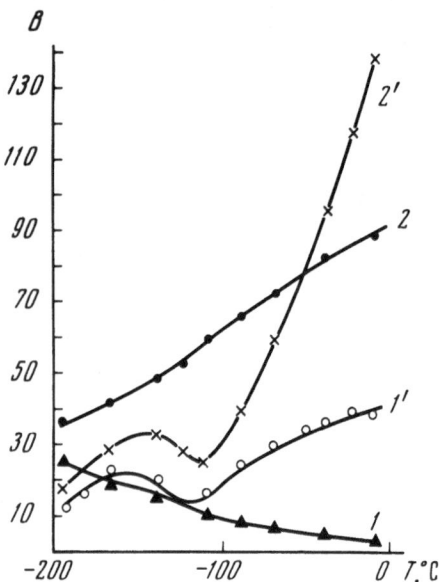

Fig. 9. Temperature dependences of the brightness of the photoluminescence (1, 2) and electroluminescence (1', 2') bands: 1), 1') blue band; 2), 2') green band.

by T. I. Voznesenskaya at the Luminescence Laboratory of the Physics Institute of the Academy of Sciences of the USSR, Moscow). The experimental results obtained by L. A. Vinokurov and kindly supplied to the author are presented in Fig. 9. Vinokurov determined the temperature dependences of the brightness of the electroluminescence and photoluminescence emitted by phosphor No. 142(3). He found that the temperature dependences of the brightness of the blue and green photoluminescence bands were monotonic, whereas the corresponding dependences of the electroluminescence had a maximum and minimum for each of these two bands. A comparison of the temperature dependences of the electroluminescence and photoluminescence brightness also supported our initial conclusion. If the extrema in the B = B(T) dependence were due to a redistribution of holes between two levels of ionized activator centers (resulting from a change in the equilibrium conditions with increasing temperature), this should have affected both the electroluminescence and photoluminescence. This was not observed. Consequently, one should seek the cause of the nonmonotonic temperature dependence of the electroluminescence brightness in the special nature of the excitation of activator centers in electroluminescence. This conclusion was supported also by other experimental data which will be considered later.

We have mentioned earlier a shift of the temperature dependence of the brightness along the T axis when the frequency of the excitation voltage was increased (Fig. 5).

A graphical analysis of the relationship between the frequency of the external field and the reciprocal of the temperature corresponding to the minimum of the electroluminescence brightness yields a straight line, shown in Fig. 10. This linear relationship is

$$\log \nu = C_1 - C_2 \frac{10^{-3}}{T},\tag{4}$$

where C_1 and C_2 are constants. As mentioned earlier, the electroluminescence brightness beyond the minimum increases proportionally to the rectified current and in the range of temperatures where the brightness decreases this current is constant. Figure 11 shows the dependence, on the external field frequency, of the reciprocal of the temperature at which the plateau changes to a rise. This dependence is also given by Eq. (4) and the values of C_1 and C_2 are the same as for the line in Fig. 10. It is known from [8, 10] that the value of the rectified current is governed by the processes which occur in the depletion barriers during that half-period when the electroluminescent phosphor is excited. The resultant brightness is affected not only by the excitation process but also by the migration carriers and their subsequent recombination. Our analysis shows clearly that the characteristics of the excitation process are reflected most clearly in the brightness at 200°K < T < 270°K. Thus, the key to the observed brightness rise is the behavior of the rectified current. Unfortunately, the temperature dependence of the rectified current in this range of temperatures cannot yet be explained. However, we can make some tentative comments on this subject.

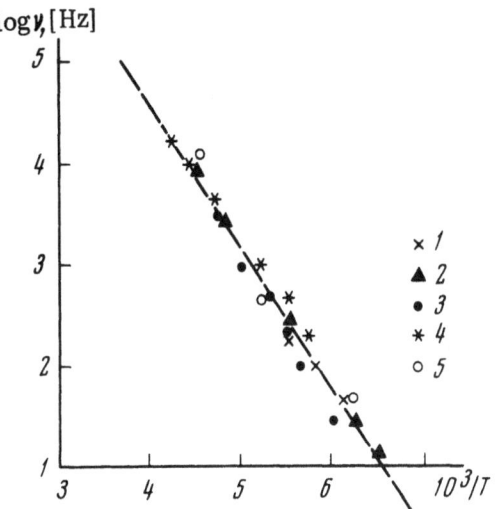

Fig. 10. Dependence of the frequency of the external field on the temperature corresponding to the electroluminescence brightness minimum. Samples emitting green luminescence: P-204 (1), P-212 (2), P-207 (3), R-308 (4), P-2 (5).

The lines plotted in Figs. 10 and 11 can be used to determine the constants C_1 and C_2 and the corresponding constants A and Δ in the following equations:

$$\nu = A \exp\left(-\frac{\Delta}{kT}\right). \qquad (5)$$

The activation energy Δ is thus found to be 0.28 $\pm$ 0.03 eV; log A = C_1 = 12 $\pm$ 1. According to some authors [3, 5], the constants A and Δ determined in this way represent the frequency factor and the energy depth of traps. The depths of traps in ZnS:Cu calculated in this way cover a wide range of values: 0.15 and 0.2 eV [4], 0.25 eV [12], 0.57 and 0.7 eV [3]. The determinations of the depth of traps in copper-activated zinc sulfide carried out by two different but reliable methods have given 0.1 and 0.35 eV [14], 0.21, 0.24, 0.26, and 0.34 eV [15]. Our value of the activation energy (Δ = 0.28 $\pm$ 0.3 eV) is in good agreement with some of these values but this does not prove that Δ is the energy depth of traps.

It is even more difficult to accept the interpretation of the role of postulated traps which is suggested in [3, 5]. It is assumed there that traps accumulate electrons transferred by the field from the cathode region of a crystal and that Eq. (5) gives the probability of liberation of electrons during the recombination half-period of the applied field. It is difficult to accept this explanation because such a process should affect equally the electroluminescence brightness and the rectified current, whereas the rectified current characteristics are sensitive only to the excitation. Consequently, if we assume that the activation energy Δ represents the depth of traps, it follows that the liberation of electrons should occur during the excitation half-period.

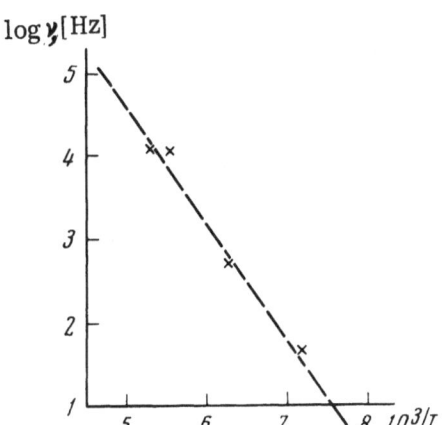

Fig. 11. Dependence of the frequency of the external field on the temperature, deduced from the rectified current characteristics.

Since the probability of thermal excitation of traps whose depth is 0.3 eV is negligible at the temperatures considered, we may assume that carriers are liberated from the traps by the field but the tunnel liberation process begins at higher temperatures than the leakage through the 0.3 eV surface barrier [8] because such tunnel liberation occurs without the participation of the host-lattice phonons.

Even greater objections can be raised against the interpretation of A as the frequency factor of traps. The published value of the frequency factor of traps in ZnS:Cu is 10^8 sec^{-1} [15] and the constant A found in [4, 11] on the basis of Eq. (5) is 10^{10} sec^{-1}. The value obtained in the present paper is A $\approx$ 10^{12} sec^{-1}. Moreover, this value can be even higher, as demonstrated by an analysis of the dependence whown in Fig. 12. This figure gives the frequency of the external field as a function of the reciprocal of the tempe-

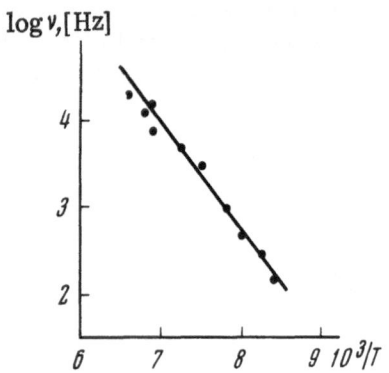

Fig. 12. Dependence of the frequency of the external field on the temperature corresponding to the maximum of the electroluminescence brightness. Sample P-212.

rature corresponding to the electroluminescence brightness maximum. The points in Fig. 12 fit very well a straight line although the scatter is greater than for the brightness minimum. The line in Fig. 12 gives $A = 10^{15}$ sec^{-1} and $\Delta = 0.28$ eV (within the limits of the experimental error). All this forces us to the conclusion that the proposed interpretation of the parameter A is doubtful and that a more reliable proof of its physical nature is required.

It must be pointed out that the temperature dependence of the electroluminescence brightness found in the present investigation is typical not only of the ÉL-510 phosphors emitting green luminescence but also of other electroluminescent phosphors based on ZnS. Figure 13 shows the experimentally obtained temperature dependences of the brightness for samples emitting blue electroluminescence (ÉL-460 phosphor) when excited with U = 100 V at a frequency of 12 kHz. A comparison of the results obtained for different electroluminescent phosphors shows that the nature of the phosphor affects the slope of the falling part of the dependence and of the subsequent rise, which is observed when the temperature is increased. The differences in the slopes are particularly noticeable in this region. For example, the slope of log B = $f(1/T)$ for the ÉL-510 phosphor is 350-400°K and the slope for ÉL-460 ranges from 700 to 850°K.

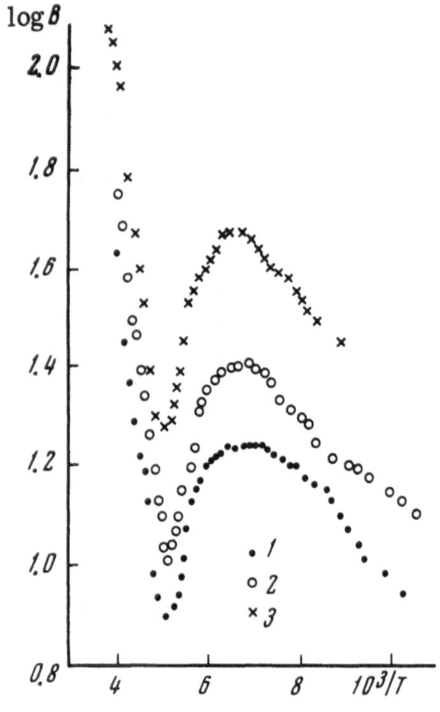

The slope of the rising part of the dependence located at the lowest temperatures is practically the same (100 ± 20°K) for both phosphors. This has been checked on more than 20 samples exhibiting nonmonotonic dependences B = B(T). Excitation fields of different intensities and frequencies have been used in this check. No such similarity is found for samples exhibiting a monotonic rise of the brightness. The slope of the low-temperature region changes considerably not only with the nature of the activator in ZnS but also with the amplitude of the external field. This is indicated by the results tabulated in [3]. This factor is probably responsible for the very low value of the phonon energy deduced in [6] from the low-temperature region of the monotonically rising curve B = B(T). Since the low-temperature slope of those B = B(T) curves which have maxima and minima is independent of the nature of the activator and the excitation conditions, we may assume that this slope is related to the properties of the host substance, such as the energy of the lattice phonon. If the phonon is generated in tunnel transitions, we can use Eq. (1) and the results given in [8] to find the phonon energy which is within the 0.03-0.045 eV range for different phosphors used in electroluminescent panels. This energy is very close to the energy of optical phonons in ZnS. More-

Fig. 13. Temperature dependence of the brightness for samples emitting blue electroluminescence: P-117 (1), P-118 (2), P-119 (3).

over, we must mention that the slopes of the low-temperature regions of the dependences $B = B(T)$ and $\overline{\mathscr{J}} = \overline{\mathscr{J}}$ (T) may be equal within the limits of the experimental error (sample P-26) or the slope of $\overline{\mathscr{J}}$ may be twice as large as the slope of B (sample P-251). This provides indirect evidence of the influence of the tunnel effect on the low-temperature electroluminescence brightness because, in this range of temperatures, the rectified current $\overline{\mathscr{J}}$ is determined by tunnel transitions assisted by one or two optical phonons of the host lattice.

Thus, an analysis of the temperature dependence of the brightness gives some (but no more than that) information on the mechanisms which determine the brightness. The physical meaning of the parameters which can be deduced from the temperature dependence of the brightness is not clear and great caution should be exercised in the interpretation of these parameters.

Literature Cited

1. I. K. Vereshchagin, Opt. Spektrosk., 16:290 (1964).
2. A. N. Georgobiani and M.V. Fok, Opt. Spektrosk., 10:188 (1961).
3. C. H. Haake, J. Electrochem. Soc., 104:291 (1957).
4. V. V. Nuvar'eva and P.E. Ramazanov, Izv. Vyssh. Ucheb. Zaved., Fizika, No. 3, p. 113 (1963).
5. G. F. Alfrey, Brit. J. Appl. Phys., 6:Suppl. No. 4 (Luminescence), p. S44 (1955); J.B. Taylor, ibid., p. S45.
6. A. N. Georgobiani, E.Yu. L'vova, and M.V. Fok, Opt. Spektrosk., 15:95 (1963).
7. Yu. P. Chukova and M.V. Fok, Zh. Tekh. Fiz., 35:762 (1965).
8. M. V. Fok and Yu.P. Chukova, Zh. Tekh. Fiz., 35:1139 (1965).
9. L. V. Keldysh, Zh. Eksp. Teor. Fiz., 34:962 (1958).
10. Yu. P. Chukova, Tr. Fiz. Inst. Akad. Nauk SSSR, 37:149 (1966).
11. I. K. Vereshchagin, Opt. Spektrosk., 16:651 (1964).
12. P. D. Johnson, W.W. Piper, and F.E. Williams, J. Electrochem. Soc., 103:221 (1956).
13. M. V. Fok and Yu. P. Chukova, Zh. Tekh. Fiz., 35:2065 (1965).
14. V. V. Antonov-Romanovskii and L.A. Vinokurov, Opt. Spektrosk., 1:71 (1956).
15. H. Gobrecht and D. Hofmann, J. Phys. Chem. Solids, 27:509 (1966).
16. J. T. Randall and M.H.F. Wilkins, Proc. Roy. Soc., London, A184:366 (1945).

PRINCIPLES OF CONVERSION
OF ELECTRICAL ENERGY INTO LIGHT

Yu. N. Nikolaev and M. V. Fok

The physical basis for increasing the efficiency of thermal and electroluminescent light sources is considered. Calculations are given of the optimal parameters of a semiconductor to be used as an incandescent body and as an electroluminescent light source. These parameters include the forbidden band width, the optimal impurity concentration, and the depth of impurity levels. It is shown that the replacement of tungsten with a plate of a suitably selected semiconducting material should double or treble the light output of an incandescent lamp. It is also shown that the efficiency of injection-electroluminescence light sources may approach 100%.

Introduction

About 10% of the total electrical power output is used in the form of lighting. And yet the efficiency of light sources remains very low. Thus, for example, an incandescent lamp converts less than one-tenth of the consumed power into visible light. A fluorescent lamp has a higher efficiency but even it converts only about a quarter of the power input into light. In spite of much work which has been done on light sources, the progress in the last few decades has been relatively slow. This seems to suggest that the limits of efficiency have been reached in the existing methods for the conversion of electrical energy into light and that a search should be made for new methods.

However, before we seek these new methods, we must make sure that they exist in principle. In other words, we must determine whether the low efficiency of current light sources is not limited by some thermodynamic aspects in the same way as the efficiency of a steam engine. We can easily show that there is no such limitation. Although a thermodynamic limit to the efficiency of light sources does exist, it lies well above the efficiencies which have been achieved so far. A static electric field has zero entropy. Light has a nonzero entropy if it is not monochromatic and not formed into a strictly parallel beam. (The entropy of light is governed by the spectral density of radiation traveling along a given direction.) The creation of entropy by the emission of light may compensate the annihilation of entropy caused by the cooling of a luminous body, provided, of course, such cooling is not too great. Thus, thermodynamic reasoning shows that a luminous body may convert into light not only all the electrical power supplied to it but also some of its intrinsic thermal energy. This problem has been considered quantitatively by Weinstein [1]. He came to the conclusion that the thermodynamic limit of the efficiency of the conversion of electrical energy into light, i.e., the ratio of the emitted radiant energy to the consumed electrical energy, is

$$\eta = \frac{T_{\text{eff}}}{T_{\text{eff}} - T} , \qquad\qquad (1.1)$$

where T_{eff} is the effective radiation temperature*; T is the temperature of the luminous body.

Weinstein also presented estimates showing that the value of η may appreciably exceed unity, even for fairly high radiation powers.

Thus, we are very far from reaching the thermodynamic limit of efficiency in the conversion of electrical energy into light. We must find ways of approaching this theoretical limit. We shall now consider this problem in more detail.

Sources of light can be divided into thermal and luminescent (fluorescent). Thermal sources of light — incandescent bodies — operate at temperatures well above the ambient temperature. Therefore, the second law of thermodynamics predicts that they can lose heat only to the ambient medium (by conduction, convection, and radiation). If a thermal source of light is placed in vacuum, the conduction and convection losses are practically eliminated. If, moreover, the conditions are such that the radiation is emitted only in the visible part of the spectrum, all the electrical energy supplied to an incandescent body (this energy is used simply to maintain the high temperature of this body) is converted into light. Thus, the thermodynamic limit of the efficiency of thermal light sources is 100%. The real efficiency is far lower.

Luminescent light sources remain cold during the emission of light. In principle, they may be colder than the ambient medium. Then, heat may flow from the ambient medium into the luminous body and this may provide the necessary (but not sufficient) condition for the efficiency to exceed 100%, as predicted by Weinstein.

We shall now consider the physical principles which can be used to approach the thermodynamic limit of the efficiency of thermal and luminescent light sources. We shall bear in mind that although currently the efficiency of luminescent light sources is somewhat higher than that of thermal sources, it is *a priori* not clear which type of source is capable of further development and which of them will be most practical. It is very likely that, as found in current practice, both thermal and luminescent sources will find their own areas of usefulness.

§1. Thermal Light Sources

The low light output of modern incandescent lamps is due to the fact that most of the radiation emitted by a tungsten filament is concentrated in the infrared rather than in the visible range of wavelengths. This is because the absorptivity of tungsten varies very little with the wavelength and, therefore, Kirchhoff's law predicts that the spectrum of its radiation in the incandescent state should be close to the emission spectrum of an absolute blackbody. According to Wien's law, the maximum in the emission spectrum of an absolute blackbody shifts in the direction of shorter wavelengths when the temperature is increased. Therefore, the fraction of the visible radiation in the spectrum of an absolute blackbody should increase with increasing temperature provided the latter is not too high. The greatest proportion of the visible radiation in the spectrum of an absolute blackbody is observed at a temperature of about 7000°K. When the temperature is increased still further, the visible fraction of the radiation begins to decrease because of the rapid increase in the amount of emitted ultraviolet radiation. However, no known substance remains solid at 7000°K. Moreover, even if we were able to construct a black emitter

---

*This is the temperature of an absolute blackbody emitting radiation which carries away the same amount of entropy.

operating at this temperature (for example, a plasma filament), the visible radiation in this most favorable case would represent only about 37% of the total emitted power.*

The situation would be different if the luminous element were transparent in the infrared region. Then, according to Kirchhoff's law, it would emit no radiation at infrared wavelengths so that all the emission would be concentrated in the visible (and partly in the ultraviolet) range of wavelengths.

The light output of a lamp with such a luminous element would be very high. However, the problem is to find a suitable substance which would be transparent at infrared wavelengths.

Fok [2] demonstrated that a thermally stable semiconductor with a wide forbidden band is such a substance.

The following parameters of a semiconductor are of importance in light-source applications:

1) the forbidden band width (energy gap) E_G;
2) the temperature dependence of E_G (it is usually assumed that E_G decreases linearly with increasing temperature in accordance with $E_{G,T'} = E_{G,O} - \beta T$, where $\beta = (3-6) \times 10^{-4}$ eV/deg);
3) the impurity concentration N;
4) the depth E of the impurity energy level (if $E \ll E_G$, i.e., if the impurity level is sufficiently close to one of the allowed bands, it may be assumed that E is independent of T);
5) the lattice vibration frequencies (they govern the positions of the absorption bands in the infrared region);
6) the mobility μ of free carriers and its temperature dependence;
7) the effective mass m^* of charge carriers;
8) the nature of the fundamental absorption edge (direct or indirect, allowed or forbidden transitions).

We shall now consider how these properties are related to the operation of a semiconductor incandescent lamp and, particularly, to its efficiency. The efficiency will be defined as the ratio of the energy emitted in the visible part of the spectrum to the total energy supplied to the lamp. If an incandescent semiconductor is kept in vacuum, the energy is emitted only in the form of radiation. Therefore, in this case, the total energy supplied can be replaced by the total energy emitted as radiation.

According to Kirchhoff's law, the power of the radiation emitted by an incandescent body in a frequency interval $d\nu$ is

$$\varepsilon(\nu, T)d\nu = K(\nu, T)A(\nu, T)d\nu, \tag{1.2}$$

where $A(\nu, T)$ is the emissivity of an absolute blackbody (Planck's function) and $K(\nu, T)$ is the absorptivity of the incandescent body. The power of the visible radiation W_{vis} is

$$W_{vis} = \int_{\nu_{IR}}^{\nu_{UV}} \varepsilon(\nu, T)\, d\nu, \tag{1.3}$$

---

*We shall define the visible radiation as that lying between photon energies of 3 and 1.8 eV. The maximum of the sensitivity of the human eye lies in the middle of this range. The sensitivity at 3 and 1.8 eV is only 0.5% of the maximum value. Outside these limits the sensitivity of the eye is even less and can be ignored completely.

where $h\nu_{IR} = 1.8$ eV ($\lambda = 685$ nm) and $h\nu_{UV} = 3$ eV ($\lambda = 412$ nm) are the limits of the visible range, as defined in the present investigation.

The total radiation power W_{rad} is

$$W_{rad} = \int_0^\infty \varepsilon(\nu, T)\, d\nu. \tag{1.4}$$

The efficiency η of an incandescent body in vacuum (not heat conduction) is

$$\eta(T) = \frac{W_{vis}}{W_{rad}}. \tag{1.5}$$

Using Eqs. (1.2)–(1.5), we find that

$$\eta(T) = \frac{\displaystyle\int_{\nu_{IR}}^{\nu_{UV}} K(\nu, T)\, A(\nu, T)\, d\nu}{\displaystyle\int_0^\infty K(\nu, T)\, A(\nu, T)\, d\nu}. \tag{1.6}$$

More rigorously, we should calculate not the radiant efficiency but the luminous efficiency φ using the formula

$$\varphi(T) = \frac{\displaystyle\int_0^\infty V(\nu)\, \varepsilon(\nu, T)\, d\nu}{\displaystyle\int_0^\infty \varepsilon(\nu, T)\, d\nu}, \tag{1.7}$$

where $V(\nu)$ is the luminous efficiency function (the dependence of the sensitivity of the eye on the frequency of the light incident on the eye). However, the introduction of the function $V(\nu)$ complicates the calculations quite considerably without introducing any basic changes in the results. Therefore, we shall simply calculate the radiant efficiency η.

We shall consider how the emission spectrum of a semiconductor varies when its temperature is increased. Figure 1 shows schematically the emission spectra of a semiconductor at various temperatures. This figure is plotted on the assumption that the transparency range of the semiconductor is limited by the forbidden band width and by the energy of normal lattice vibrations, and that the absorptivity at these limits changes discontinuously from unity outside the transparency region to 0.01 within that region.* Figure 1 shows clearly how much the infrared emission decreases because of the transparency in the infrared range.

The spectra obtained at different temperatures (Fig. 1) show how the visible radiation fraction varies with increasing temperature and how the transparency range becomes narrower. As long as the short-wavelength limit of the transparency range lies in the visible region, an increase in the temperature causes the visible radiation power to increase more rapidly than the total radiation power (compare Figs. 1a and 1b). The ultraviolet radiation power increases even more rapidly but its absolute value at temperatures corresponding to Figs. 1a and 1b is

---

*We shall show later that the situation is actually more complex. However, Fig. 1 represents correctly the basic features of the emission spectrum of an incandescent semiconductor.

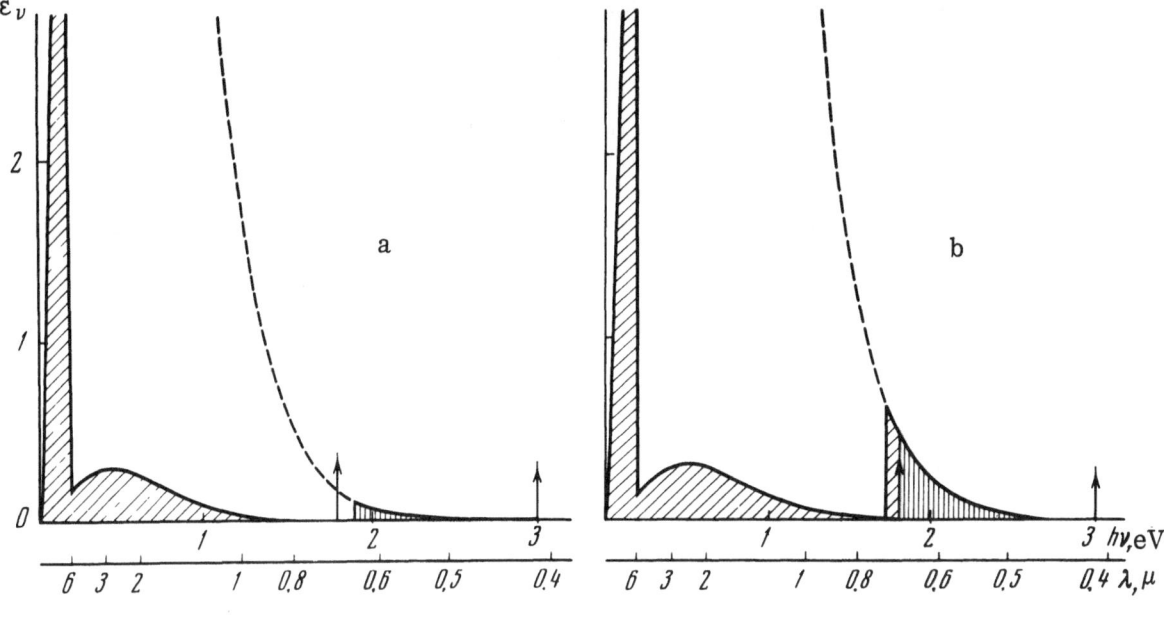

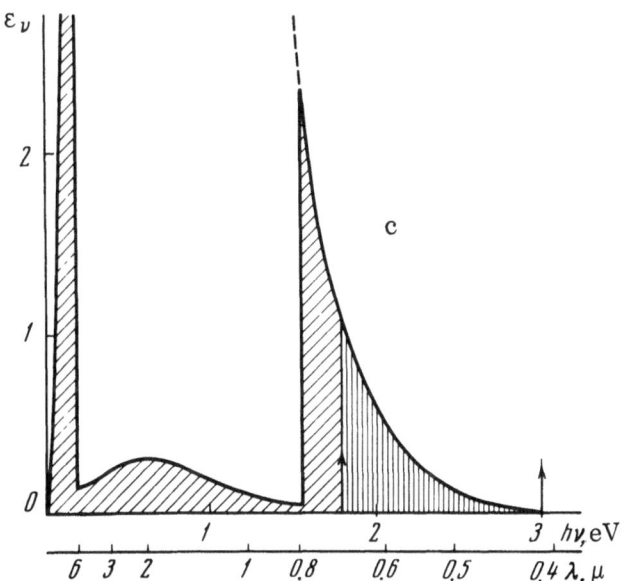

Fig. 1. Schematic representation of the
influence of temperature on the emission
spectrum of an incandescent semiconductor: 1) 1800°K; b) 2100°K; c) 2400°K.
Vertical hatching represents the visible
region and the cross hatching represents
the infrared region. The dashed curve
is the spectrum of an absolute blackbody.
The arrows indicate the limits of the
visible region.

low and such radiation is not very important. However, when the limit of the transparency range
shifts to the infrared region, it is found that the infrared radiation power increases rapidly
(compare Figs. 1b and 1c). This reduces the radiant efficiency, making it approach the efficiency of an absolute blackbody at the same temperature. Thus, the optimal working temperature of
a semiconductor incandescent lamp is lower than that of an absolute blackbody and the maximum
radiant efficiency of a semiconductor may be higher than the maximum efficiency of a blackbody,
and much higher than the efficiency of a tungsten filament. This will become clear in an analysis
of the influence of the parameters of an incandescent body on its efficiency as a source of visible light.

It is evident from Eq. (1.6) that the function $\eta(T)$ can be calculated only if we know the
absorptivity $K(\nu, T)$. The absorptivity is governed by the parameters of the incandescent body

(listed earlier) and by its geometrical shape (in the case of a thin plate, it is governed by the thickness). We shall calculate K(ν, T) making the following simplifying assumptions.

1. We shall ignore the Fresnel reflection. This cannot distort the results very greatly because such reflection varies little with the energy of the incident quanta and, therefore, it reduces to practically the same extent the visible and the total radiation output. Consequently, there is very little change in the ratio of these outputs.

2. The absorptivity in the interband (band-band transition region) will be assumed to be unity and we shall postulate that the absorptivity decreases suddenly at a photon energy equal exactly to the forbidden band width at a given temperature $E_{G,T}$. In fact, the absorption coefficient decreases smoothly when the photon energy is reduced and the steepness of the absorption edge depends on the nature of electron transitions at the wavelength of this edge. This point will be considered again and some refinements will be made.

3. We shall also assume that the absorptivity is unity at frequencies equal to or lower than the frequency of normal lattice vibrations ν_l. This overestimates somewhat the absorptivity in the infrared region since this region includes transparency and specular reflection bands. However, the transparency bands lie in the far infrared where A(ν, T) is small and the reflection bands are narrow. Therefore, the transparency and reflection bands do not alter greatly the general energy balance.

4. We shall assume that the absorptivity in the $h\nu_l < h\nu < E_{G,T}$ range is due to the absorption by free carriers. The semiconductor concerned will be posulated to contain only one principal impurity and all impurities of opposite sign will be ignored. The simplest possible formula for the free-carrier absorption will be used:

$$\alpha(\nu) = \frac{ne^3}{4\pi c m^{*2} n_r \nu^2 \mu} , \qquad (1.8)$$

where e is the electronic charge; n is the free carrier density; m^* is the effective mass of carriers (assumed to be equal to the free-electron mass); μ is the mobility, which will be assumed to be independent of temperature; c is the velocity of light; ν is the frequency; n_r is the refractive index. This formula overestimates somewhat the absorption coefficient, particularly at low values of the frequency ν and of the mobility μ. Consequently, the efficiency is underestimated.

5. We shall assume that the absorptivity is proportional to the absorption coefficient α (this is rigorously true only at low values of α and overestimates the absorptivity at high values of the absorption coefficient*) up to ν_m, at which $\alpha = 100$ cm^{-1}. At lower values of ν, for which the coefficient of absorption by free electrons is even higher, the absorptivity will be assumed to be unity.

Thus, according to our assumptions,

$$\begin{aligned}
K(\nu, T) &= 1 \quad &\text{for} \quad &E_{G,T} \leqslant h\nu; \\
K(\nu, T) &= \alpha \quad &\text{for} \quad &\left.\begin{array}{r} h\nu_l \\ h\nu_m \end{array}\right\} < h\nu < E_{G,T}; \\
K(\nu, T) &= 1 \quad &\text{for} \quad &\text{even lower values of } h\nu.
\end{aligned} \qquad (1.9)$$

Figure 2 shows our calculated temperature dependence of the efficiency of plates of different thickness prepared from the two samples of a semiconductor† differing only in the impurity (donor) concentration. The upper curve in each pair corresponds to $N = 10^{17}$ cm^{-3} and

---

*The coefficient of proportionality in front of α is governed by the thickness of the plate.

†The parameters of this semiconductor are: $E_{G,O} = 3.0$ eV; $\beta = 5.2 \times 10^{-4}$ eV/deg; $\mu = 2$ cm$^2 \cdot$ V$^{-1} \cdot$ sec^{-1}; $h\nu_l = 0.2$ eV.

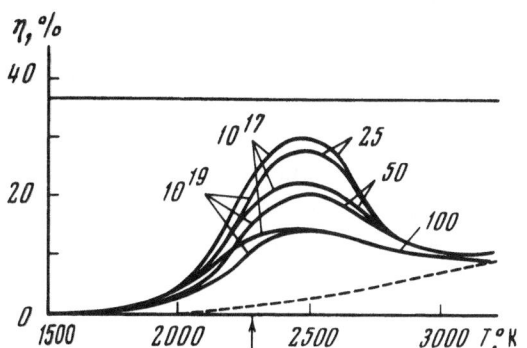

Fig. 2. Influence of the thickness of an incandescent semiconductor on the temperature dependence of its efficiency. The numbers on the right give the thickness of a semiconductor plate in microns; the numbers on the left give the impurity concentration. The vertical arrow represents the temperature at which the limit of the transparency region coincides with the boundary between the visible and infrared parts of the spectrum.

the lower to $N = 10^{19}$ cm^{-3}. The dashed curve represents the efficiency of an absolute blackbody. The horizontal line, plotted above all the other curves, represents the maximum efficiency which could, in principle, be achieved for an absolute blackbody at about 7000°K. We can see that the curves for our semiconductor lie considerably higher than the dashed curve, which represents a blackbody at temperatures below 3000°K. We can see that, at 7000°K, the efficiency of an absolute blackbody would be higher than the maximum efficiency of our semiconductor. However, no blackbody capable of operating at 7000°K has been found yet. Real incandescent elements, made of tungsten and having a spectrum close to that of an absolute blackbody, can operate up to temperatures of only 3000–3500°K. The efficiency of tungsten at these temperatures is half the efficiency of our semiconductor at 2500°K.

It is evident from Fig. 2 that an increase in the thickness of an incandescent body reduces the efficiency. This means that the semiconductor plate should be as thin as possible. Calculations show that the role played by the thickness increases as the transparency range widens in the direction of longer wavelengths.

Figure 3 shows the temperature dependence of the efficiency for various positions of the long-wavelength limit of the transparency range, imposed by the lattice vibrations, and for two impurity concentrations. The curves for $h\nu = 0.2$ eV in Figs. 3a and 3b are identical with the curves for 50 μ thickness in Fig. 2. For the sake of comparison, the same curves are used again in Figs. 4 and 5. We can see that when the transparency range expands in the direction of longer wavelengths, the maximum efficiency increases and the temperature at which this

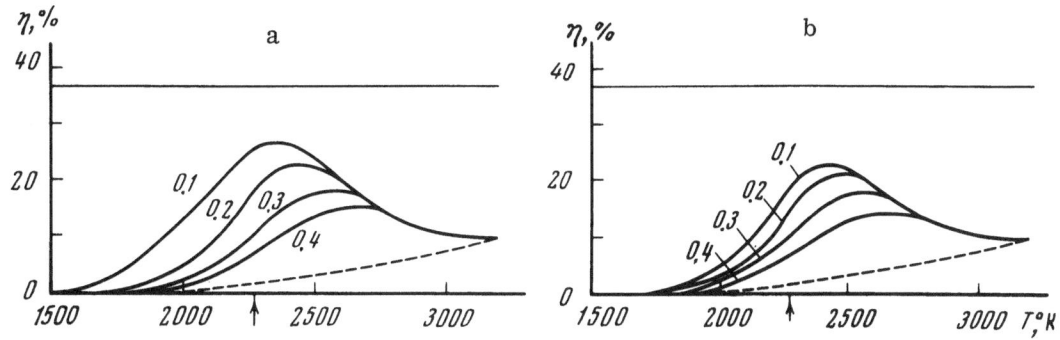

Fig. 3. Influence of the position of the long-wavelength limit of the transparency region on the temperature dependence of the efficiency of an incandescent semiconductor. The numbers alongside the curves give the value of $h\nu_l$ in eV. Impurity concentration: a) $N = 10^{17}$ cm^{-3}; b) 10^{19} cm^{-3}. The horizontal line represents the maximum possible efficiency of a blackbody at about 7000°K.

maximum occurs decreases. Calculations show that these variations increase with decreasing absorptivity in the transparency range. It is also evident from Fig. 3 that a considerable gain in the efficiency is achieved if the semiconductor begins to absorb in the infrared region not at 6 μ (0.2 eV) but at 4 μ (0.3 eV).

The results presented in Figs. 2 and 3 have a simple physical meaning. An increase in the absorptivity in the transparency region and the narrowing of this region from the long-wavelength side result in an increase in the emission of the infrared radiation and thus make the emission spectrum of the semiconductor resemble more closely the spectrum of an absolute blackbody. It is evident from these figures that an increase in the infrared absorptivity makes the efficiency curves of the semiconductor approach the efficiency curve of an absolute black-body.

Figure 4 gives the corresponding curves for a semiconductor which differs only by a ten-fold increase in the mobility from the semiconductor considered so far. We can see that such an increase in the mobility considerably raises the efficiency and makes the maximum significantly narrower. Under some conditions, the efficiency of the new semiconductor may even exceed the efficiency of an absolute blackbody at 7000°K. It is interesting to note that, in all cases, the efficiency maximum appears when the short-wavelength transparency edge shifts to the infrared region.

Figure 5 demonstrates the influence of the forbidden band width of a semiconductor and of its temperature coefficient on the temperature dependence of the efficiency. Our calculations are based on the assumption that the long-wavelength transparency limit, imposed by the lattice vibrations, lies at 6.2 μ (hν_l = 0.2 eV) and that the thickness of the semiconductor plate is 50μ. The vertical "columns" of graphs in Fig. 5 represent semiconductors which differ only in their forbidden band width $E_{G,O}$; the horizontal "rows" of graphs represent semiconductors which differ only in the temperature coefficient β of their forbidden band width. The uppermost curve in each set represents a semiconductor with a free-carrier mobility μ = 20 cm$^2 \cdot$ V$^{-1} \cdot$ sec^{-1}. The two middle curves represent a semiconductor with μ = 2 cm$^2 \cdot$ V$^{-1} \cdot$ sec^{-1} (they differ in the impurity concentration: the upper one applies to 10^{17} cm^{-3} and the lower to 10^{19} cm^{-3}). The lowest (dashed) curve represents the temperature dependence of the efficiency of an absolute blackbody.

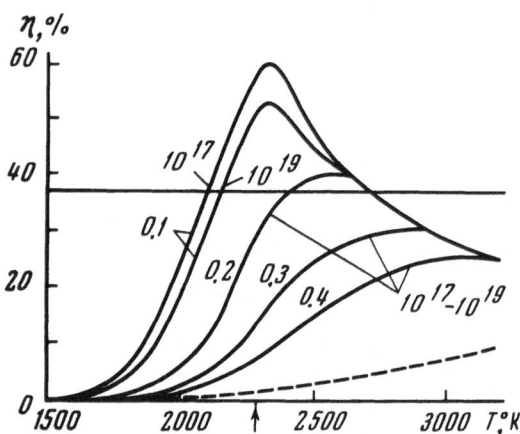

Fig. 4. Temperature dependence of the efficiency of an incandescent semiconductor with a carrier mobility of 20 cm$^2 \cdot$ V$^{-1} \cdot$ sec^{-1}. The numbers alongside the curves have the same meaning as in Figs. 2 and 3.

We can see from Fig. 5 that, when the forbidden band width increases, the efficiency maximum shifts in the direction of higher temperatures and becomes narrower, and that the maximum efficiency increases. However, at fairly low temperatures, an increase in the forbidden band width may reduce the efficiency. It follows that there is an optimal forbidden band width for each working temperature (other conditions are assumed to be constant). This can be explained as follows. If the forbidden band is very wide, the short-wavelength edge of the transparency region does not reach the boundary separating the visible and infrared parts of the spectrum at the working temperature (Fig. 1a). Consequently, the visible radiation power decreases whereas the infrared radiation power remains constant. If the forbidden band is very narrow, strong absorption extends — at the work-

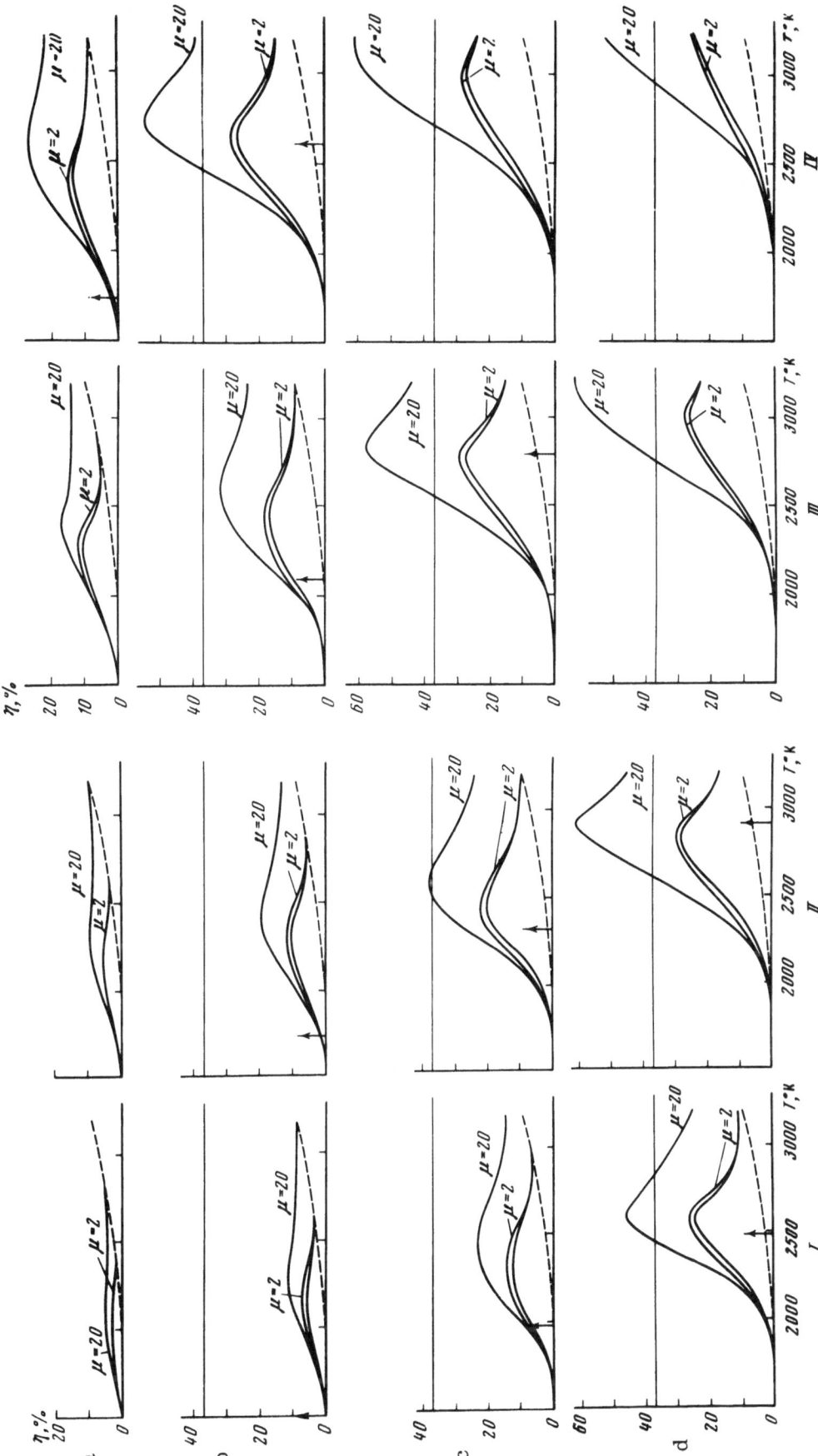

Fig. 5. Influence of the forbidden band width $E_{G,O}$ and its temperature coefficient β on the temperature dependence of the efficiency of an incandescent semiconductor with $E_{G,O} = 2.4$ (a), 2.7 (b), 3.0 (c), and 3.3 eV (d) and $\beta = 6 \times 10^{-4}$ (I), 5.2×10^{-4} (II), 4.3×10^{-4} (III), and 3.4×10^{-4} eV/deg K (IV).

ing temperature — far into the infrared region (Fig. 1c). This enhances the infrared output leaving the visible radiation unchanged. In both cases, the ratio of the visible to the infrared radiation power decreases, i.e., the efficiency diminishes. The same result is produces by an increase in the free-carrier absorption. Therefore, in some cases, the efficiency begins to decrease even before the short-wavelength transparency edge reaches the infrared region (the temperature at which this occurs is indicated in Fig. 5 by an arrow, whenever this arrow can be fitted within the range of the plotted curve).

The effect of absorption by free charge carriers may be lessened by reducing the thickness of the semiconductor plate. However, this may also reduce the absorptivity in the fundamental absorption region due to the interband transitions. Such a reduction may occur, for example, if these transitions are indirect or forbidden, since in these cases the absorption is weak (of the same order as the free-electron absorption). Therefore, it would be desirable to use a semiconductor with a fundamental absorption edge due to direct allowed transitions (in this case, the edge is steepest).

The influence of the coefficient β on the temperature dependence of the efficiency can be deduced by referring to Fig. 5 and comparing the graphs in the same horizontal "row." It is found that a reduction in β has the same result as an increase in $E_{G,O}$. Thus, if $E_{G,O}$ is constant, there is an optimum value of β for each working temperature. The explanation of this observation is the same as that in the case of the forbidden band width itself.

Thus, a favorable combination of the values of the forbidden band width and of its temperature coefficient, and reasonable values of the other parameters of a semiconductor, can ensure a very high efficiency at 2400–2600°K, which may even exceed the efficiency of an absolute blackbody at its optimal temperature of 6500–7000°K.

The required combination of properties can be achieved experimentally because we can select not only the chemical composition of a semiconductor but also its crystal modification with a suitable forbidden band width. For example, SiC is known to exist in several tens of hexagonal polytypes whose forbidden band widths differ by almost 1 eV.*

Other compounds may also be suitable. They include some rare-earth sulfides, titanium and thorium dioxides, and other substances. They have not yet been investigated from the point of view of their possible applications in semiconductor lamps.

The introduction of special impurities may exert a very considerable influence on the absorptivity spectrum of a semiconductor. The presence of impurities may affect the efficiency if the absorption by these impurities lies in the transparency region. An impurity absorption band located in the visible range near the absorption edge of the host substance will be shifted by heating in approximately the same way (though usually more slowly) as that absorption edge. In the optical sense, this is equivalent to a reduction in the forbidden band width with all its consequences. Thus, the introduction of impurities can be useful if other conditions are suitable.

If an impurity absorption band lies in the infrared region, the presence of the corresponding impurity can only reduce the efficiency. It follows that such impurities are undesirable from the optical point of view but the presence of some impurities is nevertheless desirable from the point of view of the improvement of the electrical properties of the incandescent body.

_____

*Kauer [3] is of the opinion that SiC is unsuitable for this purpose because its free-carrier absorption is far too high. However, Kauer does not give experimental data on the infrared absorption in SiC at high temperatures. Calculations carried out by the present author using Kauer's formula suggest that the free-electron absorption at high temperatures should not be too large. Only careful experiments can resolve this contradiction.

The position of an impurity absorption band is usually governed by the position of the impurity energy level in the forbidden band of the host semiconductor. For example, if the level of a donor impurity is located close to the valence band, the absorption due to this impurity occurs close to the absorption edge of the host semiconductor. If this level is shallow, i.e., if it is located close to the conduction band, the associated optical absorption band lies in the near or far infrared. Calculations indicate that impurities, whose absorption bands are located at wavelengths longer than 9-10 μ, reduce only slightly the efficiency of a semiconductor thermal source of light. The most harmful are those impurities which have donor levels located at a depth of 0.5-1 eV from the conduction band edge since the associated absorption occurs in that part of the infrared region in which the emission of an absolute blackbody is particularly strong. All these remarks about donor levels apply also to acceptor levels except that the level position should be measured from the top of the valence band.

We shall now consider briefly the influence of the carrier mobility on the efficiency of an incandescent semiconductor. It is evident from Fig. 5 that an increase in the mobility at the working temperature results in an appreciable increase of the efficiency. This is due to the reduction in the absorption by free carriers. Therefore, one should give preference to covalent semiconducting materials because the mobility in such materials is usually higher than that in other semiconductors.

We shall now discuss the electrical properties of an incandescent body.

The heating of a tungsten filament by an electric current presents no difficulties. The situation is different in the case of a semiconductor plate. The conductivity of a semiconductor increases when its temperature is increased. This raises the current flowing through the plate and increases the Joule heat dissipation, which raises the temperature. This increases the current again and the whole cycle is repeated. It follows that a current-heated semiconductor may become unstable.

A ballast resistance, connected in series with a semiconductor, is needed to stabilize the heating. In this sense, the passage of a current through a semiconductor resembles a discharge in a gas because of the need for a ballast resistance to stabilize the current. The presence of a ballast choke is one of the most important disadvantages of fluorescent units.

This difficulty can be avoided in the case of semiconductor lamps. Calculations show that, under certain conditions, a current-heated semiconductor may remain stable without any ballast resistance (in spite of the rapid increase of its conductivity during its heating). These conditions are achieved if the rate of dissipation of heat is greater than the rate of rise of temperature. Then, an increase in the Joule heat due to an accidental rise in the temperature cannot maintain the higher temperature because the heat is dissipated too quickly.

The temperature dependence of the total radiation power is given by Eq. (1.4). The conditions of stable heating in vacuum can be found from the temperature dependence of the Joule heat W_J. If the applied voltage is constant, the Joule heat is proportional to the electrical conductivity. Thus, the temperature dependence of the relative value of W_J can be found by calculating the temperature dependence of the electrical conductivity $1/\rho$. The following assumptions will be made in this calculation.

1. We shall assume that our semiconductor is doped with impurities which are completely ionized at temperatures of interest in our application, i.e , above 1500°K. The impurity concentration (or, more exactly, the difference between the donor and acceptor concentrations) will be denoted by N.

2. The electron and hole mobilities will be taken as equal and their effective masses will be assumed to be equal to the free-electron mass.

3. The carrier mobility will be assumed to be inversely proportional to the absolute temperature. This assumption is usually justified at high temperatures.

4. The thermal and optical widths of the forbidden band will be assumed to be equal.

All these assumptions simplify considerably our calculations without altering the basic features of the dependences being considered. More accurate calculations can be carried out if the high-temperature parameters of a semiconductor are known.

Thus, we shall assume that

$$\frac{1}{\rho} \propto \frac{1}{T} \sqrt{N^2 + 4n_i^2}. \qquad (1.10)$$

The factor $1/T$ appears because of the temperature dependence of the mobility. The value of n_i is given by

$$n_i = 4.82 \cdot 10^{15} T^{3/2} \exp\left(-\frac{E_{G,0}}{2kT} + \frac{\beta}{2k}\right) \text{cm}^{-3}, \qquad (1.11)$$

where T is the temperature in deg K.

At high temperatures, an incandescent body loses heat mainly by radiation. Figure 6 shows schematically the temperature dependence of the radiation power W_{rad}. The curve denoted by W_J represents the temperature dependence of the Joule heat evolved during the passage of a current in the case when N is so small that the conduction remains intrinsic throughout the range of temperatures considered. The point of intersection of the W_{rad} and W_J curves is an equilibrium state corresponding to the equality of the Joule heat and the power emitted as radiation. To the left of this point, the Joule heat is greater than the radiation power. This means that, at temperatures below the equilibrium point, the incandescent body is heated by the electric current. On the right of the intersection point, the W_J curve lies below W_{rad}. This means that the radiation power is greater than the Joule heat and, therefore, the body cools in spite of the increasing heat generated by the electric current. In both cases, the temperature of the incandescent body varies in such a way as to reach the equilibrium point. Therefore, the equilibrium state is stable. Conversely, if heat is lost mainly by conduction, the high-temperature equilibrium is unstable. This also follows from Fig. 6. The curve designated by W_c represents the temperature dependence of the heat loss by conduction. The thermal conductivity is selected so that the W_c curve intersects the W_J dependence at the same point as W_{rad}. We can see that the W_c curve lies above W_J on the left and below it on the right. This means that if the temperature is below the equilibrium value, the loss of heat will exceed the Joule heat and the body will cool. If the temperature is above the equilibrium value, the losses will exceed the amount of heat supplied by the Joule effect and the temperature of the body will rise. In both cases,

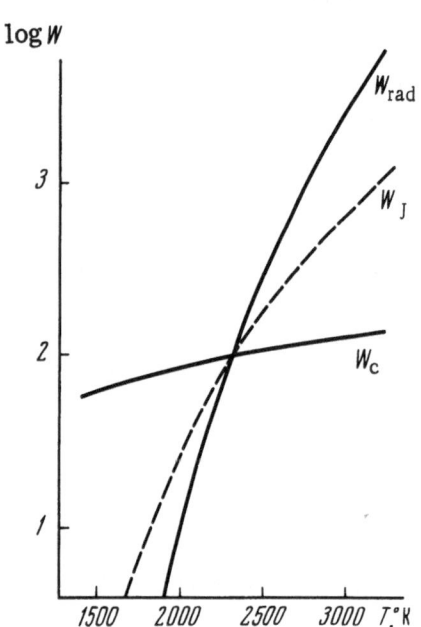

Fig. 6. Schematic representation of the stability of the heating conditions in a semiconductor. W_J is the power consumption (Joule heat); W_{rad} is the power lost as radiation; W_c is the power lost by conduction.

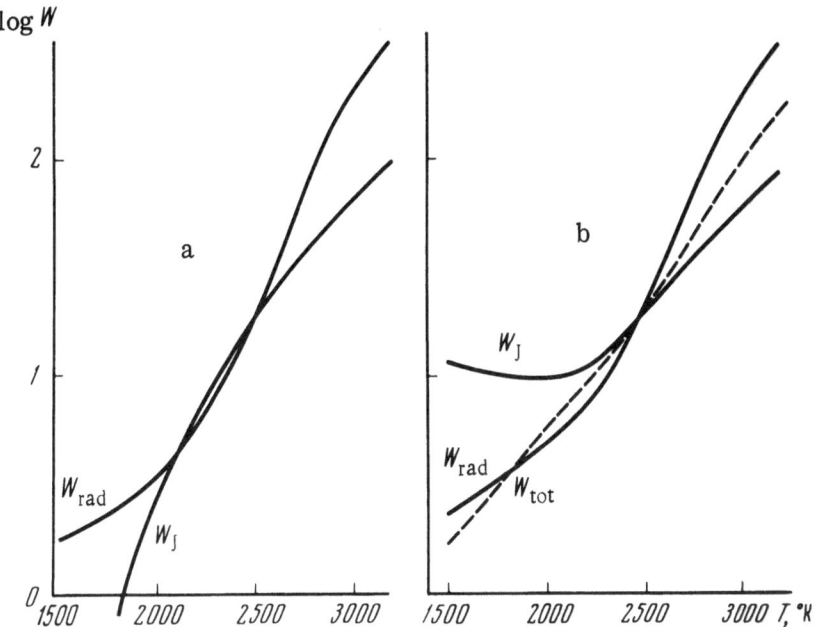

Fig. 7. Stability of the heating conditions in a semiconductor.
W_J is the Joule heat; W_{rad} is the power lost as radiation; W_{tot}
is the total lost power. All the curves are made to cross at
$T = 2500°K$ by moving along the ordinate. Impurity concentration:
a) 10^{17} cm^{-3}; b) 10^{19} cm^{-3}.

the conditions will tend to drift away from the equilibrium state, i. e., this state is unstable.

Thus, a semiconductor heated by an electric current is in a stable equilibrium if the heat is lost mainly by radiation. Happily, the requirements for stability and high efficiency are identical. It follows that if we wish to achieve a high efficiency we must ensure that the energy supplied is converted mainly into light and not dissipated by heat conduction.

The value of W_{rad} in Fig. 6 is calculated for a semiconductor which is perfectly transparent throughout the infrared region and, therefore, is identical with W_{vis}. In the case of a real semiconductor, the W_{rad} curve is much less steep and in some cases its slope is even less than that of W_J. This happens because real semiconductors always have finite absorption throughout the infrared region. Therefore, the infrared emission makes a considerable contribution to the radiation energy balance but this contribution increases (with rising temperature) much more slowly than the radiation in the visible and ultraviolet regions.

If a semiconductor contains many impurities, its temperature dependence of W_J is less steep than that shown in Fig. 6. It follows that, once again, a stable state may be achieved. Figure 7 shows the temperature dependences of W_J for two semiconductor* samples with donor impurity concentrations differing by a factor of 100. We can see that an increase in the impurity concentration (Fig. 7b) produces an almost horizontal region in W_J curve between 1500 and 2100°K. The minimum in this curve is due to the fact that all the donors become ionized at some temperature. Therefore, further heating does not increase the number of electrons in the conduction band but it reduces the mobility and, therefore, the conductivity. (In Fig. 7a, this region lies at lower temperatures than those used in the graph.) When the temperature is in-

---

*The temperature dependence of the efficiency of an incandescent body made from this semi-
conductor is shown in Fig. 2.

creased still further, intrinsic conduction begins to play a role, i.e., the density of the electrons in the conduction band increases because of the transitions from the valence band. This results in a rapid increase in the electrical conductivity.

The W_{rad} curves in Fig. 7 represent the temperature dependences of the radiation power. They correspond to the $W_{rad} = W_{vis}$ in Fig. 6 but include the power emitted in the form of visible, infrared, and ultraviolet radiation. The W_{rad} curves intersect W_J at about 2500°K and the equilibrium at this point is stable since the W_{rad} curve is, at this point, steeper than the W_J curve. The W_J curves are calculated for a voltage selected to ensure their intersection with W_{rad} at 2500°K.

If the voltage is increased by a factor m, the Joule heat increases by a factor m^2 and the corresponding curves are shifted upward by $2 \log m$ without a change in their shape (the scale along the ordinate is logarithmic). The point of intersection of the W_J and W_{rad} curves naturally shifts to a different temperature. It is evident from Fig. 7b that the W_{rad} curve is always steeper than W_J. This means that if an incandescent body is made from a semiconductor with a given impurity concentration (10^{19} cm^{-3}), an adjustment of the applied voltage is all that is required to achieve any desired temperature and to maintain it for a long time. On the other hand, the W_J curve in Fig. 7a is steeper than W_{rad} up to 2300°K. This means that a semiconductor with an impurity concentration of 10^{17} cm^{-3} cannot remain stable in this range of temperatures. At 2300°K, the vapor pressure of such a semiconductor may be so high that the semiconductor will begin to dissociate. It follows that the donor concentration in this particular semiconductor should exceed 10^{17} cm^{-3}. The depth of the donor levels has almost no influence on the stability of the equilibrium since all the donors are ionized well before the working temperature is reached. For example, levels located at a depth of 0.22 eV (nitrogen in SiC) are almost completely ionized at temperatures slightly higher than room temperature.

We have considered so far the operation of a semiconductor incandescent lamp in vacuum. This was the lamp considered by Kauer [3]. However, it is very likely that the service life of a semiconductor lamp may be increased by operating it in an inclosure filled with an inert gas, as is done in the case of tungsten lamps. In this case, heat will be lost not only by radiation but also by conduction and by convection, the latter loss being proportional to the temperature drop between the incandescent body and the enclosing bulb. For simplicity, we shall refer to the conduction and convection losses simply as the losses. As the temperature is raised these losses increase much more slowly than the radiation output. In an ordinary (tungsten) incandescent lamp, the losses represent about 25% of the total power supplied. In a semiconductor lamp, these losses should be of the same order of magnitude because the geometrical conditions are similar but the actual numerical value of the losses may be greater because less power will be required for a given light output. The presence of these losses makes the heat removal rate W_{tot} (we can no longer call it W_{rad} because it now includes conduction and convection losses) less steep than W_{rad}. The temperature dependence of W_{tot} for N = 10^{19} cm^{-3}, represented by the dashed curve in Fig. 7b, is calculated on the following assumptions. The losses at 2500°K are assumed to represent 0.25 of the power emitted by an absolute blackbody at the same temperature. This corresponds approximately to the losses in a tungsten incandescent lamp. It is also assumed that the losses are proportional to the temperature difference between the incandescent body and the ambient medium surrounding the lamp (this medium is assumed to be at 300°K). We can see that the W_{tot} curve is also steeper than the W_J curve but the difference between their slopes is considerably less than the difference between the slopes of W_{rad} and W_J. It follows that even a slight reduction in the dopant concentration may disturb the stability of the heating conditions.* On the other hand, an increase in the impurity concentration is undesirable because it

_____

*It is understood that a ballast resistance can always be used to stabilize the heating conditions but such a resistance complicates the construction of the lamp and is, therefore, undesirable.

may reduce the transparency in the infrared region as a result of the increase in the free-electron absorption.

Our calculations apply to a semiconductor specified by its parameters but the results apply also to other real semiconductors which can be used as an incandescent body. This is because the forbidden band widths of suitable semiconductors should be approximately the same at the working temperature (these widths are governed by the red edge of the visible region). It follows that the temperature dependences of the conductivities of suitable semiconductors should not differ greatly and, consequently, the desirable impurity concentration should be of the same order of magnitude for all suitable materials. The impurity need not be of the donor type. If the electron and hole mobilities are similar, the same results can be obtained using an acceptor impurity with a level at the same depth as that of the donor.

We shall consider another problem associated with the operation of a semiconductor incandescent lamp. It is known that the electrical conductivity of semiconductors decreases rapidly when the temperature is lowered. Therefore, the initial conductivity of a semiconductor (at room or even freezing temperature) may be so low that the current flowing through it at the working voltage will be insufficient to raise the semiconductor to the required temperature. It would then be necessary to use an additional heater or initially to apply a higher voltage. In both cases, a special starting device would be needed.

Calculations show that these difficulties can be avoided using impurities which produce relatively shallow levels. For a semiconductor with $E_{G,O} = 3$ eV, the maximum depth is 0.3–0.4 eV, depending on the initial temperature, provided the impurity concentration is sufficient to establish stable heating conditions at high temperatures. Figure 8 shows the temperature dependences of the Joule heat and the power dissipated by radiation, conduction, and convection throughout the range of temperatures of interest between the initial and working temperatures. Curves 1 and 2 represent a semiconductor with 10^{19} cm^{-3} donors, whose levels are located at depths of 0.12 and 0.3 eV, respectively. These curves merge at high temperatures. Curves 3 and 4 represent the temperature dependences of the total power dissipation for lamps located in ambient media kept at 200°K (–73°C) and 250°K (–23°C), respectively. We can see that if the donor levels are located at 0.12 eV, the Joule heat exceeds the dissipation of heat at both temperatures but if the level depth is 0.3 eV, this happens only at 250°K. This means that, in all the cases considered (and also at low initial temperatures if the donor depth is 0.3 eV), a semi-

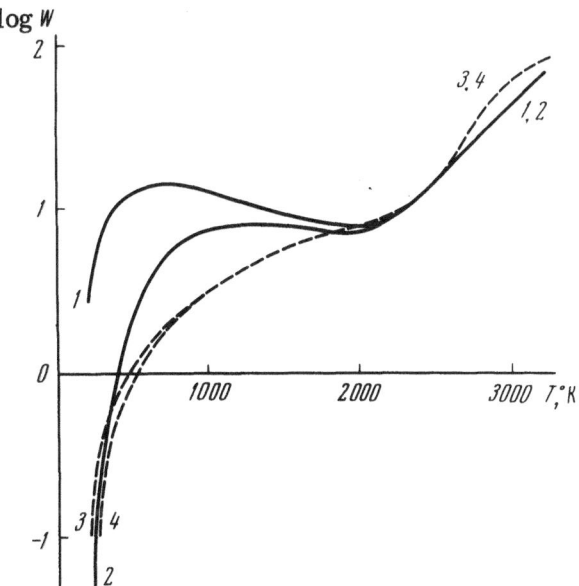

Fig. 8. Starting conditions of a semiconductor incandescent lamp at low temperatures: 1) the Joule heat for donors with a level at a depth of 0.12 eV; 2) the same as before but for 0.3 eV; 3) the power lost for an initial temperature of –73°C; 4) the power lost for an initial temperature of –23°C.

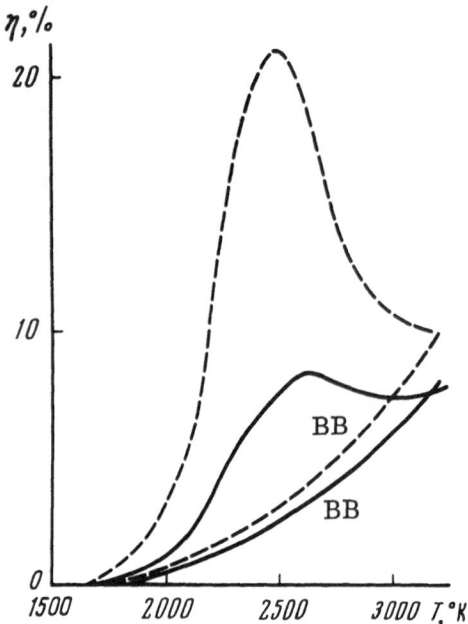

Fig. 9. The influence of heat conduction and convection losses on the efficiency of a semiconductor incandescent lamp. The dashed curves represent the calculations made without allowance for these losses; the continuous
curves include the losses. The scale
along the ordinate is 5 times that in
Fig. 5.

conductor may be heated electrically to any required
temperature without additional starting devices.

It follows that if the semiconductor is doped
with a suitable impurity, no starting devices are
required. If at all temperatures (except the working
temperature) the Joule heat is considerably greater
than the dissipated power, a semiconductor lamp will
heat up rapidly (the rate of heating will be slightly
less than that of a tungsten lamp because the maximum current does not flow immediately through the
semiconductor). This can achieved by the use of
sufficiently shallow donors or acceptors.

We shall now consider the problem of efficiency. We have mentioned earlier that the heat conduction and convection losses in a semiconductor incandescent lamp would be higher than those in a tungsten
lamp. The higher the relative luminous efficiency of
a semiconductor lamp, the lower would be the losses
through unwanted radiation and, consequently, the
higher would be the fraction of energy carried away
by conduction and convection. Figure 9 shows the
temperature dependences of the efficiency of a semiconductor lamp and a conventional tungsten lamp.
The dashed curves in Fig. 9 are plotted ignoring the
losses, and the continuous ones are plotted making
allowance for the losses. We can see that the conduction and convection losses reduce the efficiency
of a semiconductor lamp much more than the effi-
ciency of a tungsten lamp. Nevertheless, at temperatures below 3200°K, the efficiency of a
semiconductor lamp remains higher than that of a tungsten lamp (at the same working temperature). The maximum efficiency of a semiconductor lamp (reached at 2600°K) is slightly higher than that of a tungsten lamp (reached at 3200°K). These curves are plotted for a semiconduc-
tor with $E_{G,O}$ = 3 eV, β = 5.2 × 10^{-4} eV/deg K, and μ = 2 cm^2 · V^{-1} · sec^{-1}. Naturally, the efficiency of a semiconductor lamp would be higher for μ = 20 cm^2 · V^{-1} · sec^{-1}.

The relative luminous efficiency of a gas-filled semiconductor lamp can be appreciably
increased by enclosing it in a double-walled bulb on the Dewar flask principle. The inner wall
of such a bulb would be kept at a higher temperature than that of a conventional tungsten lamp,
which has no vacuum jacket. This would reduce the temperature difference between the incandescent semiconductor and the outer wall of the bulb. Such a reduction would also decrease the
conduction and convection losses.

Since the inner wall of the bulb would be separated from the ambient space by a vacuum,
heat could be lost from this wall solely by radiation (this would be infrared radiation because
the bulb must be transparent in the visible region). Calculations show that a glass bulb * of
5 cm diameter, kept at a temperature of 600°K and surrounded by a vacuum jacket, would ra-
diate about 40 W, i.e., about the same amount as a bulb of a semiconductor lamp equivalent to
a 100-W tungsten lamp. Usually, the temperature of a bulb of an incandescent lamp is about
400°K. An increase in this temperature by 200 deg K would reduce the heat conduction and
convection losses by 10% and correspondingly increase the relative luminous efficiency by 6%

*We shall assume that glass is transparent up to λ = 2.5 μ.

above the value which could be obtained without a vacuum jacket (the heat conduction and convection losses of a semiconductor lamp are estimated as 60% of the total power consumption). Quartz is transparent at long infrared wavelengths (up to $\lambda = 4.5\ \mu$) and, therefore, a quartz bulb of the same size as a 100-W tungsten lamp would have a temperature of 850°K if surrounded by a vacuum jacket. This means that the use of a vacuum jacket would increase the relative luminous efficiency by 16%. If the bulb size were reduced to about half that of a 100-W tungsten lamp, without reducing the power, the luminous efficiency would be doubled. Thus, a vacuum jacket would be worthwhile in the case of small high-power lamps.

We shall now formulate briefly the principal conclusions on the subject of thermal light sources.

1. A semiconductor with a forbidden band width of 2.5-3 eV, capable of operating at 2000-2500°K, would be a suitable material for an incandescent lamp. This semiconductor should be transparent up to 4-6 μ.

2. Such a semiconductor should be doped with impurities giving rise to donor (acceptor) levels of depths not exceeding 0.3 eV. It would be desirable to use an impurity which increases the density of those carriers whose mobility is higher. Stable heating conditions require an impurity concentration of the order of 10^{19} cm^{-3}.

3. An incandescent body should be a crystalline plate as thin as possible (its thickness should not exceed 0.1 mm).* The length of this plate would be governed by the applied voltage and its width by the required luminous flux.

A semiconductor incandescent lamp satisfying these requirements should have service characteristics close to those of a conventional tungsten lamp except for its efficiency, which should be double or treble that of a tungsten lamp.

§2. Luminescent Light Sources

Electroluminescent light sources can be divided into two large groups: gas-discharge sources, in which the electroluminescence of a gas is used in some way, and solid-state sources, which are based on the electroluminescence of crytal phosphors. The main energy losses in the luminescent light sources are due to the conversion of some of the energy supplied into heat (within the source itself), and to the emission of wavelengths outside the visible part of the spectrum.

We shall consider first the gas-discharge sources operating at low pressures and low current densities. In these sources, luminescence is generated by "hot" electrons which collide inelastically with the atoms of the gas filling the discharge tube. These collisions excite or ionize the gas atoms. The electrons are heated by absorbing energy from an electric field. The gas ions and atoms in the tube remain relatively "cold." This is because neutral atoms are incapable of acquiring energy directly from an electric field and ions move so slowly (their masses are tens of thousands of times greater than the electron mass) that they do not experience a sufficiently large potential drop between two successive collisions. Consequently, the thermal losses in such light sources are very low. In some cases, the efficiency of these sources reaches 80%. The excited atoms give up their energy in the form of light quanta of a definite frequency because the radiating atoms hardly interact with other atoms. Therefore, the emission spectra of such sources consist of isolated and very narrow spectral lines. If a tube of this kind is filled with a mixture of various gases or if several light sources emitting different colors are placed in a mat bulb, a very efficient source of white light is obtained.

---

*On this point, we differ from Kauer [3]. He regards 0.2 mm as the minimum permissible thickness since he considers only an evacuated lamp in which the evaporation of the incandescent body would be very rapid.

However, since the emission spectra of such sources are basically of the line type, the colors of various objects viewed in the light of these sources are different from those observed in sunlight. For this reason, the use of these lamps is very limited: they are normally employed in illumination of intercity highways and in "neon-type" outdoor advertising.

Other disadvantages of the gas-discharge light sources are their relatively low brightness and their relatively high working voltages.

Two methods are used to improve the color rendition of these sources. In one method, the electroluminescence of a gas is not used directly but it is employed to excite the photoluminescence of some specially selected phosphor with a broad emission spectrum. This method is employed in the widely used fluorescent lamps. However, the two-stage conversion results in considerable losses. This is because the quantum efficiency of the photoluminescence does not exceed unity. This means that each quantum absorbed by a phosphor can give rise to only one quantum of luminescence. Moreover, according to the Stokes law, the energy of the emitted quantum will be less than that of the exciting quantum. The best color rendition is given by white light containing quanta of all the wavelengths in the visible range. Therefore, white light can be emitted by a phosphor if it is excited by quanta lying at the short-wavelength end of the visible range. At least one-third of the energy is lost in the conversion of these quanta into white light. In practice, the losses are even greater because the phosphors in currently used fluorescent lamps are excited with the mercury line at 2537 Å, which lies in the far ultraviolet. Therefore, over half of the energy of the gas discharge is lost in its conversion to visible white light.

The other way to improve color rendition is to increase the pressure in the electroluminescent gas and to increase the current density. This broadens the lines and produces a background between them. The result is a much improved color rendition. However, since the relatively dense plasma has a high thermal conductivity, over half of the energy supplied is lost as heat. A considerable fraction of the emitted radiation lies in the infrared and ultraviolet parts of the spectrum and this increases the losses still further.

When the gas pressure is increased sufficiently, we can obtain emission with a spectrum close to that of an absolute blackbody kept at a temperature of several thousand degrees. In this case, only about one-third of the total radiant energy is emitted as visible light.

This shows that white light is expensive to produce. However, from the thermodynamic point of view, white light should be obtainable for a lesser expenditure of energy than the corresponding line spectrum (of the same power) because light with a broad spectrum has a considerably higher entropy. Such reasoning shows that the disadvantage of gas discharge lamps is that the electrical energy is first converted into the energy of "hot" electrons, which results in a considerable increase in the entropy. Since the entropy can only increase, the emission of light (particularly if it has a line spectrum) is accompanied by the conversion of some of the energy of the "hot" electrons into heat. The requirement of visible light with a wide spectrum and, consequently, a high entropy is not satisfied well by gas-discharge lamps.

Solid-state electroluminescent sources of light are still in the development stage. In those light sources which are based on powdered phosphors dispersed in dielectrics (Destriau cells), the excitation of "hot" electrons is still an intermediate step. The energy losses in such light sources are of the same kind as those in gas-discharge lamps. In respect of efficiency these Destriau-type light sources can compete only with incandescent lamps.

However, there is a different class of electroluminescent light sources which are, in principle, free of this disadvantage: they are called injection light sources or Losev (Lossew) elements. These light sources are p-n or p-i-n junctions subjected to forward voltages. An electric field sets electrons and holes in motion in opposite directions and they recombine in the junction region to give off light concentrated in some part of the spectrum. The wavelengths emitted depend on the nature of the semiconductor and of the impurities present in it.

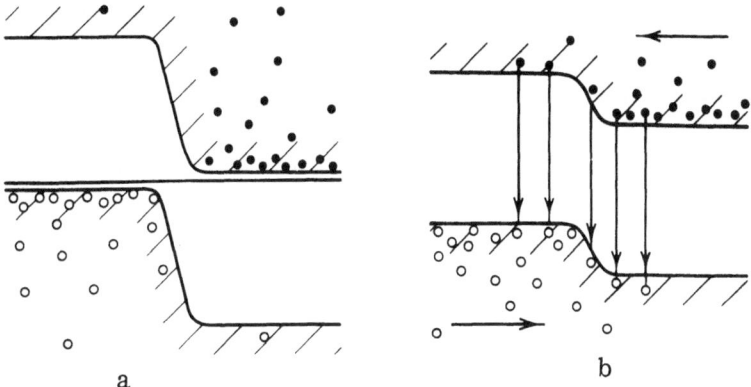

Fig. 10. Energy band scheme of a p—n junction: a) without an
external voltage; b) under a forward bias.

Figure 10 shows a simplified energy band scheme of such a junction. The p-type region is pictured on the left and the n-type region on the right. The black dots are electrons and the open circles are holes. The vertical arrows show electron transitions during recombination and the horizontal arrows represent the motion of electrons and holes. We can see that recombination takes place only between those electrons and holes which have acquired a sufficiently high energy from thermal fluctuations. The applied potential difference U reduces the barrier in the p—n junction region (compare Figs. 10a and 10b) and facilitates the transfer of electrons and holes across the junction. In contrast to the gas-discharge lamp, electrons and holes acquire (from the electric field) only some of the required energy; the rest of the energy is provided by the thermal fluctuations in the semiconductor. This results in the cooling of the semiconductor if the energy evolved during recombination is lost from the semiconductor in the form, for example, of radiation. The cooling effect at a junction of two different conductors carrying and electric current is the well-known negative Peltier effect.

The acquisition of energy from thermal fluctuations is demonstrated experimentally by the emission of appreciable electroluminescence from a p—n junction at voltages lower than the values corresponding to the energy of recombination radiation quanta (even if it is assumed that electrons suffer no energy losses in their motion) [4]. If the thermal component of the acquired energy is large, the electroluminescence brightness is low. This is because only the electrons belonging to the high–energy part of the Fermi distribution can participate in recombination processes. When the applied voltage is increased, the brightness rises rapidly and the efficiency decreases because the "gratuitous" thermal component of the acquired energy decreases. When the applied voltage exceeds the total contact potential difference,[*] no thermal component is necessary. Under these conditions, the luminescence brightness may be very high. Sometimes it is so high that a transition to stimulated emission may take place (this is the basis on which semiconductor lasers are operated).

The presence of a thermal component of the energy in each emitted quantum of light does not by itself ensure a high efficiency of electroluminescence. It is also necessary to ensure that all the recombination events result in the emission of light quanta.

Many phosphors are now available whose quantum efficiency under optical excitation is close to unity. This means that almost every recombination event results in the emission of light. Such phosphors are very attractive as electroluminescent sources into which free carriers

---

[*]This difference consists of the potential difference between one electrode and the p-type region, between the p- and n-type regions, and between the n-type region and the second electrode.

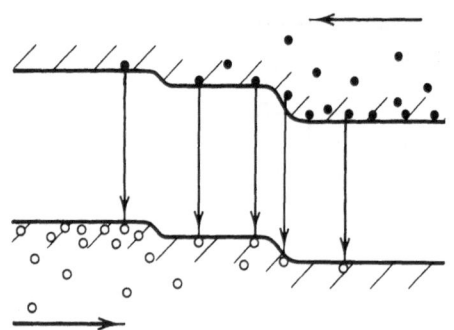

Fig. 11. Energy band scheme of a p−π−n junction subjected to a forward bias.

are injected by an electric current. This approach has been justified experimentally by the generation of infrared electroluminescence in Ge, Si, and GaAs [5].

However, in contrast to these semiconductors, the phosphors which have a high quantum efficiency in the visible part of the spectrum are usually characterized by a high resistivity because their forbidden bands are wide and because they contain comparable amounts of donor and acceptor impurities. If follows that if these phosphors are used as p- or n-type gions of p−n junctions, the Joule heat losses become large even when the current is low. Consequently, it is necessary to use a three-layer structure and place a thin phosphor layer between low-resistivity p- and n-type regions made of the same semiconductor. These outer regions have low luminous efficiencies. Therefore, it is necessary to ensure that almost all the recombination events take place in the phosphor layer. The phosphor may be excited by double injection in the p−π−n junction (the π-type region is the phosphor layer). The Joule losses will be small if the thickness of the high-resistivity π-type region does not exceed the diffusion lengths of the carriers in this region because the electrons and the holes will be distributed uniformly as a result of their diffuse motion.*

Figure 11 shows the energy band scheme of a p−π−n junction subjected to a low voltage. It is evident from this figure that recombination takes place mainly in the π-type region because only a few electrons can overcome the barrier between the π- and p-type regions (equally, only a few holes can overcome the barrier between the π- and n-type regions). The electric field is practically nonexistent in the π-type region.† Therefore, the conditions in the phosphor differ little from those obtaining in the case of photoluminescence. This will be used in later calculations.

We shall consider the recombination of electrons and holes during the excitation of a p−π−n junction. Obviously, recombination may occur in the π-type layer, at the boundaries of this layer with the p- and n-type regions, as well as within these regions. This will be true particularly when the applied voltage is sufficiently high to reduce the potential barriers between all three regions. Since we are assuming that the π-type layer is made of an efficient phosphor (we shall consider later how this can be achieved), it is desirable to enhance the recombination in this layer. On the other hand, the recombination which occurs at the boundaries between the three regions and in the p- and n-type regions is frequencly nonradiative and should, therefore, be avoided. Let us assume that the thickness of the π-type region is

$$l_\pi \lesssim L_{n\pi}, \; L_{p\pi}, \hspace{4cm} (2.1)$$

where $L_{n\pi}$ and $L_{p\pi}$ are, respectively, the diffusion lengths of the electrons and the holes in the π-type region when the phosphor is excited. We shall assume that the thicknesses of the p- and the n-type regions are considerably larger than the diffusion lengths of the minority carriers in these regions. Then, the minority carriers will not cross the whole crystal and the current will be governed entirely by recombination:

---

*The diffusion lengths of the carriers in an excited phosphor differ considerably fron the equilibrium lengths. We are speaking here of the nonequilibrium values of the diffusion lengths.
†Real diffused junctions always include a high-resistivity layer of considerable thickness [6], which is analogous to the π-type region considered here.

$$J = J_p + J_{p\pi} + J_\pi + J_{\pi n} + J_n, \tag{2.2}$$

where J is the density of the current flowing through the p−π−n junction; J_p, J_π, and J_n are the densities of the recombination current in the p-, π-, and n-type regions, respectively; $J_{p\pi}$ and $J_{\pi n}$ are the densities of the recombination current at the boundaries of the π-type layer.

The efficiency of an injection light source can be close to 100% if the recombination flux, i.e., the number of recombination events per unit time, in the π-type region makes the principal contribution to the recombination current

$$J \approx J_\pi \tag{2.3}$$

and, consequently,

$$J_{p\pi}, \; J_{n\pi} \ll J_\pi, \tag{2.4}$$

$$J_p, \; J_n \ll J_\pi. \tag{2.5}$$

We shall assume that the inequalities (2.4) and (2.5) are satisfied (later, we shall formulate the conditions under which these inequalities are obeyed).

We shall not calculate the recombination flux in the π-type region. In contrast to classical semiconductors, the impurities in efficient phosphors are not usually ionized at room temperature. In this sense, these impurities are "deep" whereas only the case of "shallow" impurities has been considered in the theory of semiconductors.

Shockley and Read [7] analyzed recombination in the case of strong excitation of a few deep centers and in the case of weak excitation of many deep centers. Neither of these cases is applicable to phosphors. In fact, typical concentrations of activators and coactivators are 1×10^{17}-5×10^{18} atoms/cm^3. The density of the injected carriers is usually much less than these concentrations but it is much larger than the equilibrium carrier density. Therefore, the space charge of the impurities in the π-type region plays an important role in the establishment of the ratio of the densities of the free electrons (n) and holes (p). In other words, these densities are not equal, as is usually assumed for the i-type region in a p−i−n junction [8]. However, the product of the electron and the hole densities Φ is, as in the case of a p−i−n junction, given by

$$\Phi = np = N_c N_v \exp \frac{eU - E_G}{kT}, \tag{2.6}$$

where N_c and N_v are the densities of states in the conduction and valence bands, respectively; U is the voltage applied to the junction.

The electrons and the holes localized at "deep" impurity centers may alter considerably the ratio of the free electron and hole densities when the degree of excitation of a phosphor is increased. This requires special analysis.

We shall now consider the process of recombination at the j-th donors which have a single energy level located closer to the conduction than to the valence band.* Under steady-state conditions, the number of electrons localized at a given center per unit time is equal to the number of delocalized electrons. Therefore, the probability of the capture of an electron (A_j) or a hole

---

*In this part of our treatment, the formulation of the problem and the results are basically the same as those given by Shockley and Read. This is true as long as we consider a recombination flux involving deep centers and expressed in terms of free carrier densities.

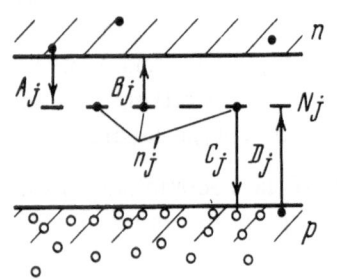

Fig. 12. Electron transitions
in an excited phosphor.

(C_j) by such a center (Fig. 12) and the probabilities of the thermal transfer of an electron from the center of the conduction band (B_j) and of the transfer of an electron from the valence band to this center (D_j) are related by the expression

$$(A_j n + D_j N_v)(N_j - n'_j) = n'_j(C_j p + B_j N_c), \qquad (2.7)$$

where N_j is the concentration of the j-th centers; n'_j is the density of the electrons localized at these centers. Hence,

$$n'_j = N_j \frac{A_j n + D_j N_v}{A_j n + C_j p + B_j N_c + D_j N_v} . \qquad (2.8)$$

The transition probabilities are related to the parameters of the recombination centers by the following equations [7]:

$$A_j = v_n \sigma_{nj}, \quad C_j = v_p \sigma_{pj}, \quad B_j = A_j \exp\left(-\frac{E_j}{kT}\right), \quad D_j = C_j \exp\left(-\frac{E_G - E_j}{kT}\right), \qquad (2.9)$$

where v_n and v_p are the thermal velocities of the electrons and holes, respectively; σ_{nj} and σ_{pj} are the electron- and hole-capture cross sections of the j-th donor; E_j is the energy of the donor level measured from the bottom of the conduction band.

The term $D_j N_v$ in Eq. (2.8) can be ignored because this term is small compared with $A_j n$ at all excitation intensities of practical interest. The dropping of this term under radiative recombination conditions implies that the thermal radiation of a given body is negligible compared with its luminescence. In this case, Eq. (2.8) becomes

$$n_j = N_j \left(1 + \gamma_j \frac{p}{n} + \frac{n_j}{n}\right)^{-1}, \qquad (2.10)$$

where

$$\gamma_j = \frac{C_j}{A_j} \approx \frac{\sigma_{pj}}{\sigma_{nj}}, \qquad (2.11)$$

$$n_j = N_c \exp\left(-\frac{E_j}{kT}\right), \qquad (2.12)$$

and n_j is the electron density in the conduction band where the Fermi level coincides with the j-th level. The recombination flux through the j-th centers is

$$I_j = n'_j C_j p = N_j \left(\frac{1}{A_j n} + \frac{1}{C_j p} + \frac{n_j}{C_j p n}\right)^{-1}. \qquad (2.13)$$

If these centers are the luminescence centers, it follows that I_j represents the light flux which appears in a unit volume of the phosphor layer. However, every phosphor contains luminescence quenching centers (recombination at these centers does not give rise to the emission of light). Moreover, a phosphor may contain impurities at which recombination is radiative but the radiation is of undesirable wavelengths.

We shall now analyze the dependence of the luminescence efficiency on the impurity content of a phosphor in a more general case* when the phosphor contains H types of different do-

*We recall that, since the charge distribution in the π-type region is practically uniform and, therefore, the electric field is practically zero, recombination in this region occurs in the same way as in photoluminescence.

nors and Q types of different acceptors. As in the case of one type of donor, we can express the density of electrons at each donor and the density of holes at each acceptor in terms of the free carrier densities n and p. We shall now write down the condition of charge conservation, which is always obeyed by uniform phosphors and which is valid in the π-type region of a p$-\pi-$n junction if the condition (2.1) is satisfied. After simplification, the condition of charge conservation becomes

$$n + \sum_{i=1}^{Q} N_i \left(1 + \frac{p}{\gamma_i n + p_i}\right)^{-1} = p + \sum_{j=1}^{H} N_j \left(1 + \frac{n}{\gamma_j p + n_j}\right)^{-1}, \tag{2.14}$$

where

$$\gamma_i = C_i / A_i, \tag{2.15}$$

$$p_i = N_v \exp\left(-\frac{E_G - E_i}{kT}\right), \tag{2.16}$$

and p_i is the density of holes in the valence band when the Fermi level coincides with the i-th acceptor level.*

The total recombination flux per unit volume of the π-type layer I_π is the sum of the fluxes through all the centers:

$$I_\pi = \sum_{i=1}^{Q} N_i \left(\frac{1}{A_i n} + \frac{1}{C_i p} + \frac{p_i}{A_i n p}\right)^{-1} + \sum_{j=1}^{H} N_j \left(\frac{1}{A_j n} + \frac{1}{C_j p} + \frac{n_j}{C_j p n}\right)^{-1}. \tag{2.17}$$

If the parameters of the centers are known, we can eliminate n and p from Eqs. (2.13), (2.14), and (2.17) and thus determine the dependence of the intensity of recombination at the j-th centers on the total recombination flux. Under optical excitation (photoluminescence) conditions, this dependence represents the dependence of the brightness on the excitation intensity (provided the j-th centers are the luminescence centers). We shall now find the conditions under which the recombination flux via the luminescence centers is the largest component of the total recombination flux, i.e., the conditions under which the quantum efficiency of the luminescence has its maximum value.

It is evident from Eq. (2.13) that, depending on the nature of the dependence of the recombination flux on the electron and hole densities, we can distinguish the following cases: second-order recombination when the last term dominates the denominator of the right-hand side of Eq. (2.13); first-order hole recombination when the penultimate term predominates; and first-order electron recombination when the first term predominates. In general, recombination is neither linear nor quadratic. Usually, one considers either first-order or second-order recombination. We cannot use this approximation and must consider the more general case of mixed recombination.

Let us assume that a static voltage is applied to a p$-\pi-$n junction. The product of the electron and the hole densities in the π-type region is Φ, and it depends only on the applied voltage in accordance with Eq. (2.6). However, if recombination at the j-th centers is not exactly a second-order process, the recombination flux through these centers will depend also on the ratio n/p. If the luminescence level is sufficiently deep (we shall denote it by subscript l) so that

$$n_l \ll n, \tag{2.18}$$

---

*In contrast to the formulas of Shockley and Read [7], Eq. (2.14) applies also in the case of strong excitation of a sample containing many impurities.

it follows from Eq. (2.13) that the recombination flux through these centers has its maximum value when $A_l n = C_l p$, i.e., when

$$\frac{n}{p} = \gamma_l \; . \tag{2.19}$$

We shall show later that n and p may differ by several orders of magnitude. The ratio of the electron to the hole density may depend on the excitation level (i.e., on Φ) because of the accumulation of free carriers at the impurity centers.

We shall try to find that impurity composition of the phosphor in the π-type layer which maximizes the radiative recombination flux at a given excitation level. In other words, a certain number of centers must be introduced into a phosphor in order to ensure that the ratio n/p is favorable for the luminescence centers and unfavorable for the quenching centers. We shall now estimate the value of γ for radiative and nonradiative recombination events. If we ignore the difference between the effective masses of the electrons and the holes, we find that the value of γ is equal to the ratio of the corresponding cross sections for the capture of free carriers, as given by Eq. (2.11).

Neutral nonradiative recombination (quenching) centers have cross sections of the order of atomic dimensions [9]

$$\sigma_{n.q} \approx 10^{-16} \, \text{cm}^2 \; , \tag{2.20}$$

and neutral radiative recombination centers have cross sections which are about four orders of magnitude smaller [10]

$$\sigma_{n.r} \approx 10^{-20} \, \text{cm}^2 \; . \tag{2.21}$$

The capture cross sections of the corresponding charged centers are governed by the field of these centers and are, therefore, the same for the nonradiative and radiative recombination centers [9]

$$\sigma_{c.q} \approx \sigma_{c.r} \approx \sigma_c \approx 10^{-14} \, \text{cm}^2. \tag{2.22}$$

Therefore, the recombination flux through the donor luminescence centers has its maximum value when

$$\frac{n}{p} \approx \frac{\sigma_{n.r}}{\sigma_c} \approx 10^{-6}, \tag{2.23}$$

and the corresponding flux through the quenching centers (which are also donors) has its maximum value when

$$\frac{n}{p} \approx \frac{\sigma_{n.q}}{\sigma_c} \approx 10^{-2}. \tag{2.24}$$

We can see that the ratios of the densities of free electrons to those of free holes which are the most favorable for the radiative and nonradiative recombination processes differ by four orders of magnitude. Therefore, we may expect that the conditions under which radiative recombination predominates over nonradiative processes may not be very rigorous. However, it does not follow that if the ratio of the carrier densities is favorable for the luminescence centers, then the recombination flux through the quenching centers will be small. The ratio of the concentrations of the luminescence and quenching centers is also important.

In practical applications it is very useful to know the concentration of the quenching centers at which a quantum radiative efficiency close to 100% can still be achieved. Let us consider this problem in more detail. Let us assume that our crystal (we are speaking here of the π-type region in a three-layer structure) contains N_l luminescence centers, which we shall assume to be donors, N_d nonradiative recombination centers (which are also donors) and N_a acceptors, at which recombination is also nonradiative. Then, it follows from Eq. (2.13) and from the analogous formula for acceptors that the recombination fluxes through these centers are

$$I_l = N_l \left(\frac{1}{A_l n} + \frac{1}{C_l p} + \frac{n_l}{C_l p n} \right)^{-1}, \tag{2.25}$$

$$I_d = N_d \left(\frac{1}{A_d n} + \frac{1}{C_d p} + \frac{n_d}{C_d p n} \right)^{-1}, \tag{2.26}$$

$$I_a = N_a \left(\frac{1}{A_a n} + \frac{1}{C_a p} + \frac{p_a}{A_a p n} \right)^{-1}. \tag{2.27}$$

We must satisfy the condition

$$I_l \gg I_d, I_a. \tag{2.28}$$

It is known that recombination is most rapid at deep centers, i.e., at centers for which the probability of the thermal ejection of an electron or a hole back to an allowed band is considerably less than the probability of their recombination. We shall assume that all three types of center are deep. The first type (denoted by l) will be taken to represent the effective luminescence centers and the other two types (represented by d and a) will be assumed to represent the most "harmful" of the quenching centers. These assumptions can be formulated as follows:

$$n_l \ll n, \quad n_d \ll n, \quad p_a \ll p. \tag{2.29}$$

We shall assume that the activator concentration is such that the number of the luminescence centers per unit volume exceeds the corresponding number of the quenching centers. Then, a phosphor should have n-type conduction and its Fermi level should lie between the level of the luminescence centers and the conduction band. This means that the density of free electrons in the phosphor will be considerably greater than the density of free holes provided the excitation level is not too high:

$$n \gg p. \tag{2.30}$$

We shall use this inequality, the inequalities in Eq. (2.29), the formula (2.12), and the estimates given in Eqs. (2.20)–(2.22) to simplify Eqs. (2.25)–(2.27), leaving only one term in the denominator of each equation. Then, the inequalities of Eq. (2.28) transform to

$$N_d \ll N_l \frac{C_l p}{C_d p} \approx N_l 10^{-4}, \tag{2.31}$$

$$N_a \ll N_l \frac{C_l p}{C_a p} \approx N_l 10^{-6}. \tag{2.32}$$

It follows that an efficient phosphor should have a very low concentration of the quenching donor and acceptor impurities. Since the concentration of the luminescence centers N_l can hardly be made much greater than 1×10^{18} cm^{-3} (at higher concentrations, the luminescence centers interact with each other and this results in quenching), it follows that the concentration of the quenching donors cannot exceed 1×10^{14} cm^{-3} and that the concentration of the quenching acceptors should be less than 1×10^{12} cm^{-3}.

Thus, the simplest method for increasing the quantum efficiency of the luminescence — an increase in the activator concentration — may not give the desired result. This may happen because, under the conditions considered, the ratio n/p may be far from the value favorable for the luminescence centers. The situation is not improved by the fact that the restriction imposed on N_a becomes less stringent when the excitation intensity is increased: the inequality of Eq. (2.30) becomes the equality and the restriction is the same as for N_d. The ratio n/p can be made to approach its optimum value by the introduction of additional shallow acceptors. If other conditions are not altered, this results in an increase in the number of free holes and a decrease in the number of free electrons because the product np depends on the applied voltage. The necessary number of acceptors can be found from the equation of electrical neutrality of an excited phosphor (2.14) by specifying the excitation level. In this case, approximately half the luminescence centers should be positively charged. We have pointed out already that the recombination cross sections of the charged centers are the same for the radiative and nonradiative case. Therefore, the quenching centers lose their "advantage" in the competition for their share of the recombination flux. Consequently, the requirements which the purity of a crystal must satisfy become less stringent.

We shall now consider this problem in a quantitative manner. We shall assume that the condition of Eq. (2.19) is satisfied by the luminescence centers, i.e.,

$$\frac{n}{p} = \frac{C_l}{A_l} = \gamma_l \approx 10^{-6}.$$

(2.33)

Then, the expressions (2.25)-(2.27) for the recombination fluxes become

$$I_l = \frac{N_l\, C_l\, p}{2 + \frac{n_l}{n}} \ ,$$

(2.34)

$$I_d = \frac{N_d C_d p}{10^4 + 10^4\, \frac{n_d}{n}} \ ,$$

(2.35)

$$I_a = \frac{N_a A_a n}{10^4 + 10^4\, \frac{p_a}{n}} \ .$$

(2.36)

These new expressions are based on the estimates (2.20)-(2.22) of the effective cross sections of the charged and neutral centers. If we drop the second term in the denominator of each of these new expressions,* we can show that the restrictions imposed by Eq. (2.28) become much less stringent [Eq. (2.28) demands that all the recombination events should occur at the luminescence centers]. Instead of Eqs. (2.31) and (2.32), we now have

$$N_d \ll N_l\ \frac{C_l p}{2 C_d p}\, 10^4 \approx \frac{1}{2}\, N_l \ ,$$

(2.37)

$$N_a \ll N_l\ \frac{C_l p}{2 A_a n}\, 10^4 \approx 50\, N_l \ .$$

(2.38)

It follows that, if the activator concentration is 1×10^{18} cm^{-3}, it is sufficient to purify a semiconductor so that the concentration of the quenching donors is 1×10^{17} cm^{-3}. We then find that

---

*We shall show later when this can be done.

90% of all the recombination events are radiative. Under these conditions, the concentration of the quenching acceptors may be of the same order of magnitude as the concentration of the luminescence centers. The recombination flux through the acceptors will be small because they become neutral when they capture electrons, whereas the luminescence centers become positive when they capture electrons.

We shall now determine the concentration of shallow acceptors N_{as} which ensures that the ratio n/p has a value favoring radiative recombination. For this purpose, we express the capture probability in Eq. (2.14) in terms of the expressions in Eq. (2.9) and we substitute into Eq. (2.14) the capture cross sections given in Eqs. (2.20)-(2.22). Making the same assumptions as in the derivation of the inequalities (2.37) and (2.38) and ignoring the density of free electrons compared with the density of free holes (because the number of electrons is 10^6 times less than the number of holes), we obtain

$$N_{as} = \left(1 + \frac{p}{p_{as}}\right)\left(\frac{N_l}{2} + N_d + p\right), \tag{2.39}$$

and hence

$$N_{as} \gtrsim \frac{N_l}{2}. \tag{2.40}$$

If the excitation level is not too high, so that

$$p \ll N_l,\ p_{as}, \tag{2.41}$$

the inequality of Eq. (2.40) can be replaced by the approximate equality.* This can be done because, in accordance with Eq. (2.37), the concentration of the quenching donors can be ignored compared with the concentration of the luminescence centers. If the concentration of the shallow acceptors is less than that given by Eq. (2.40) modified to an equality, the optimal conditions for recombination via the luminescence centers are obtained at high excitation levels. However, the concentration of the shallow acceptors cannot be too high because of the danger of concentration quenching. Therefore, optimal conditions are obtained if the concentration of the shallow acceptors is approximately equal to the concentration of the activator centers.

We can easily show that the recombination flux via the shallow acceptors will be much lower (even at maximum shallow acceptor concentrations) than the flux via the luminescence centers because these shallow acceptors will be almost completely filled with electrons.

We shall now consider the restrictions imposed by our dropping of the second terms in the expressions for the recombination fluxes. The inclusion of these terms in the expressions for I_d and I_a simply reduces somewhat the recombination fluxes via the quenching centers. This does not alter the estimates of the permissible quenching impurity concentrations. However, the second term in the denominator of Eq. (2.34) may not be small compared with the first. Consequently, the recombination flux through the luminescence centers may be considerably less than the value obtained in our calculations. In this case, the permissible concentrations of the quenching centers will be correspondingly smaller than those given by Eqs. (2.37) and (2.38). Thus, the assumption that the second term in the denominator of Eq. (2.34) is small compared with the first, imposes quite stringent limitations on the depth of the luminescence centers (this depth governs n_l).

---

*This inequality implies that the additional acceptors should be sufficiently shallow.

We thus find that

$$n_l = N_c \exp\left(-\frac{E_l}{kT}\right) < 2n = 2p\gamma_l \ . \tag{2.42}$$

Since the density of free holes should be less than N_v and the difference between N_v and N_c can be neglected, we find that the above inequality may be satisfied at room temperature if the depth of the luminescence centers is

$$E_l > 0.34 \, \text{eV} \ . \tag{2.43}$$

Thus, our calculations show that the phosphor containing many quenching impurities can have a quantum efficiency close to unity if it is doped with two impurities. The depth of the energy level of the activator centers must be at least 0.34 eV. The coactivator centers must give rise (in the ionized state) to a charge opposite to the charge of the ionized activators and the energy level of the coactivator centers should be as shallow as possible. The activator and coactivator impurities should be introduced in approximately equal amounts and the concentrations of these impurities should be as high as possible but not so high as to give rise to mutual interaction, which results in concentration quenching. The coactivator impurities make little contribution to the recombination flux but they affect strongly the ratio of the electron to the hole density in an excited phosphor and thus control the ratio of the radiative to the nonradiative recombination flux.

We shall now consider the effect of increasing the excitation level on the density of the free electrons and holes. For the sake of clarity, we shall consider the simplest possible case when a phosphor contains deep activator donors (which act as the luminescence centers) and shallow coactivator acceptors. The activator and coactivator concentrations will be assumed to be equal ($N_l = N_{as}$). We shall assume that the concentrations of other impurities are small compared with the density of the free holes throughout the range of excitation levels of practical interest. Then, we can leave only one term in each of the sums of the electrical neutrality equation (2.14). After some transformations, we obtain

$$n + \frac{N_l}{1 + \gamma_l \frac{p}{n} + \frac{nl}{n}} = p\left(1 + \frac{N_{as}}{p_{as}}\right). \tag{2.44}$$

We shall express the density of free holes in terms of the electron density, using Eq. (2.6). We shall solve the resultant equation for Φ, assuming, for the sake of simplicity, that $(1 + \frac{N_{as}}{p_{as}}) \approx 1$ (if the acceptor is sufficiently shallow, we have $N_{as} < p_{as}$):

$$\Phi = \frac{n}{2\gamma_l}\left[n(\gamma_l - 1) - n_l + \sqrt{[n(1 + \gamma_l) + n_l]^2 + 4\gamma_l n N_l}\right]. \tag{2.45}$$

We have thus obtained an equation which relates the voltage applied across a $p-\pi-n$ junction to the free electron density. This equation is obtained making use of Eq. (2.6). We can calculate the values of U and p which correspond to various specified values of n. The continuous curves in Fig. 13 show the dependences of the densities of the free holes and electrons on the applied voltage, calculated on the assumption that

$$N_l = 10^{18} \, \text{cm}^{-3}, \quad \gamma_l = 10^{-6}, \quad n_l = 10^{10} \, \text{cm}^{-3}. \tag{2.46}$$

We can see that these curves consist of several distinct regions. In each of these regions, the dependences $n(\Phi)$ and $p(\Phi)$ can be described by simpler though approximate formulas which follow from $\gamma_l \ll 1$:

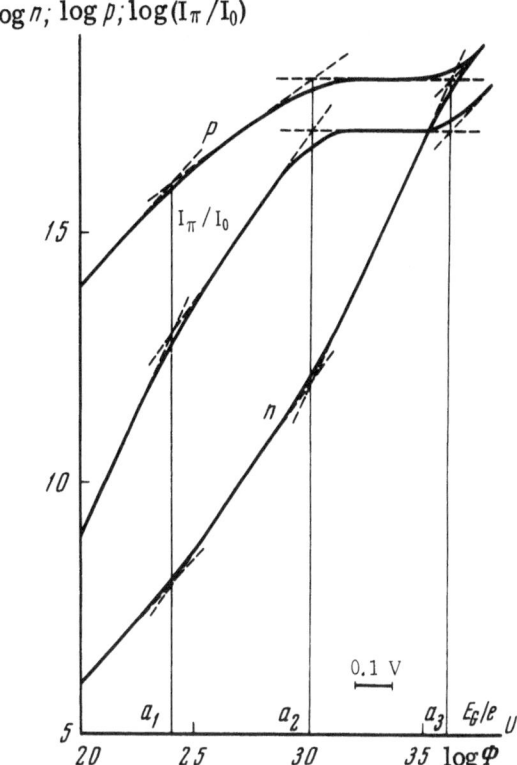

$\log n; \log p; \log(I_\pi / I_0)$

Fig. 13. Dependences of the free electron (n) and hole (p) densities and of the relative recombination flux via deep donors in the π-type region (I_π / I_0) on the voltage applied to a $p-\pi-n$ junction. The vertical lines indicate the characteristic values of U for which a_1) $\Phi = n_l^3 / N_l \gamma^2$; $p = n_l / \gamma_l$; $n = n_l^2 / \gamma_l N_l$; a_2) $\Phi = \gamma_l N_l^2$; $p = n_l / \gamma_l$; $n = \gamma_l N_l$; a_3) $\Phi = N_l^2$; $p = N_l = n$.

if $\Phi \ll \dfrac{n_l^3}{N_l \gamma_l^2}$, $n = n_l^{1/2} N_l^{-1/2} \Phi^{1/2}$,

$$p = n_l^{-1/2} N_l^{1/2} \Phi^{1/2}; \qquad (2.47)$$

if $\dfrac{n_l^3}{N_l \gamma_l^2} \ll \Phi \ll \gamma_l N_l^2$, $n = \gamma_l^{1/3} N_l^{-1/3} \Phi^{2/3}$,

$$p = \gamma^{-1/3} N_l^{1/3} \Phi^{1/3}; \qquad (2.48)$$

if $\gamma_l N_l^2 < \Phi < N_l^2$, $n = N_l^{-1} \Phi$, $p = N_l$; (2.49)

if $N_l^2 < \Phi$, $n = \Phi^{1/2}$, $p = \Phi^{1/2}$. (2.50)

Using Eq. (2.6) for Φ and the above relationships, we can easily show that in all these regions the electron and hole densities depend exponentially on the voltage:

$$n, \, p \propto \exp \frac{eU}{mkT} , \qquad (2.51)$$

where m assumes the values of 2, 1.5, 1, and 2 when the voltage is increased; the corresponding values of m for holes are 2, 3, ∞, and 2.

The dependences represented by Eq. (2.51) are shown dashed in Fig. 13. We shall now consider qualitatively the processes which determine the dependences of n and p on U in each region.

In an unexcited phosphor, all the acceptors are filled with electrons and all the donors are empty. The number of holes at the acceptors is governed, at all excitation levels, by the recombination of electrons, localized at the acceptors, with free holes and by the thermal transitions of electrons from the valence band to the acceptors. The number of localized holes is directly proportional to but much smaller than the number of free holes. Therefore, in the first approximation, all the holes in the equation for electrical neutrality can be regarded as free. Then, it is found that the sum of the densities of the free electrons and those localized at the donors is equal to the number of holes. However, in contrast to holes, most of the electrons corresponding to the first three regions in Fig. 13 are localized at impurities. In this case, Eq. (2.44) simplifies to

$$\frac{N_l}{1 + \gamma_l \dfrac{p}{n} + \dfrac{n_l}{n}} \approx p. \qquad (2.52)$$

At the lowest excitation levels [condition (2.47)], the electron population of the deep donors is governed primarily by two processes: the capture of electrons from the conduction band and their thermal transfer back to this band. Free electrons are in thermal equilibrium with those localized at the donors. This means that the densities of free and localized electrons are proportional to each other. However, the density of free holes at the lowest excitation levels (first region in Fig. 13) is equal to the density of localized electrons. It follows that the densities of the free electrons and the free holes are proportional (but not equal) to each other. The last term dominates the denominator of Eq. (2.52).

In the second region of Fig. 13, the density of holes is so high that most of the electrons at the deep donors recombine with holes and are not transferred thermally to the conduction band. Therefore, when the excitation level is increased, the density of the localized electrons increases more slowly than the density of the free electrons because an increase in the number of electron localization events is accompanied by a reduction of their lifetime in the localized state due to a rise in the probability of their recombination with free holes. Since the density of the holes is still equal to the density of the localized electrons, we find that the densities of the free electrons and holes are no longer proportional to each other. The relationship between these densities can be deduced from Eq. (2.52), in which the second term is now dominant in the denominator. It follows from the conditions in Eq. (2.48) that the second region in Fig. 13 may be observed if

$$\frac{n_l}{\gamma_l} \ll N_l \ , \tag{2.53}$$

i.e., if the concentrations of deep impurities are high.

At the boundary between the second and third regions in Fig. 13, the donors are half-filled with electrons so that a further increase in the excitation level can only double the number of electrons localized at donors. Moreover, the hole density cannot increase by more than 2, whereas the number of free electrons can increase greatly with rising voltage. This is because the probability of electron localization decreases rapidly since the number of the donors unoccupied by electrons diminishes. In this case, the denominator of Eq. (2.52) is dominated by the first term. This means that $p \approx N_l$ and hence, using Eqs. (2.5)–(2.6), we find that $n \propto \exp(eU/kT)$. At the end of the third region, the value of n becomes of the same order as p. It follows from the conditions of Eq. (2.49) that the third region is observed if

$$\gamma_l \ll 1, \quad n_l \ll N_l \ . \tag{2.54}$$

The conditions of Eq. (2.54) are important if the inequality of Eq. (2.53) is not satisfied.

In the last (fourth) region of Fig. (13), the densities of the free electrons and holes exceed the number of impurities. Therefore, the charges localized at these impurities can be ignored in the equation of electrical neutrality. In the first approximation, we may assume that $n \approx p$. The optimal concentration of the activator is usually only several times smaller than the density of states in the nearest allowed band. Therefore, the fourth region of Fig. 13 is not observed for efficient phosphors. In this respect, electroluminescent phosphors differ radically from typical semiconductors which usually satisfy n = p in the excited state.

The presence of shallow donors, which are in thermal equilibrium with the conduction band in the usual range of voltages, may alter the boundaries of the regions in Fig. 13 but not the nature of the relationships between the densities of free carriers.

All the remarks made about deep donors apply also to deep acceptors. The results obtained are applicable to any semiconductor with a forbidden band width exceeding the energy of quanta corresponding to the red edge of the visible spectrum. These results are independent of the forbidden band width because the equations contain the difference $eU - E_G$.

Knowing the dependences of n and p on the voltage, we can easily calculate the brightness– voltage and the current–voltage characteristics. We shall now mention some of the features of

these characteristics associated with the accumulation of carriers at deep impurity centers.*
If the concentration of these centers is so high that the condition (2.53) is satisfied, there is a
region in the current–voltage characteristic which satisfies

$$I \propto \exp \frac{eU}{1.5kT}. \tag{2.55}$$

Logan et al. [11] observed such characteristics in a wide range of temperatures for alloyed
gallium phosphide junctions. These characteristics have not yet been explained from the physics
point of view. These alloyed junctions have a relatively low efficiency and, therefore, we may
conclude that the main recombination flux flows through the nonradiative centers. The energies
of the quanta emitted by these junctions are close to the width of the forbidden band of gallium
phosphide, i.e., the radiative recombination centers are relatively shallow. This corresponds
to the case when the last term of the sum predominates in Eq. (2.25) and, consequently, the
intensity of electroluminescence depends on the applied voltage in accordance with the Boltzmann
formula. Therefore, the brightness–current characteristic can be described by the formula

$$R \propto \mathcal{I}^{1.5}, \tag{2.56}$$

where R is the luminescence intensity and $\mathcal{I}$ is the total current.

The experimental results are in full agreement with the above formula. If the radiative
recombination takes place at relatively deep centers, Eq. (2.25) is dominated by the first or second term of the sum. If the first term is the dominant one, the electroluminescence intensity is
directly proportional to the total current, but if the second term is dominant, we have

$$R \propto \sqrt{\mathcal{I}}. \tag{2.57}$$

Such dependences of the luminescence intensity on the current are typical of silicon carbide injection light sources [12]. The energy of the quanta emitted by these sources is considerably lower than the energy corresponding to the forbidden band width. This means that the levels
of the radiative recombination centers are deep.

Another feature of these brightness characteristics is associated with the difference between the electron–and hole–capture cross sections of the deep centers. This gives rise to a
region in the characteristics in which the recombination flux via the deep centers is almost
independent of the applied voltage (Fig. 13). The density of electrons in the π-type layer depends
on the voltage, in accordance with the Boltzmann formula. If the efficiency of a luminescent
source of light is much less than unity and the radiative recombination takes place at relatively
shallow centers, a slight increase in the current through the injection source should result in a
strong increase in the luminescence intensity. Some workers have called this the "threshold
current" effect.

However, if the quantum efficiency of electroluminescence is high, practically all the
recombination events occur at the luminescence centers and the brightness (luminescence intensity) is proportional to the current. In this case, the voltage dependence of the brightness
near the efficiency maximum can be approximated by an exponential function (this can be done
only in a narrow range of voltages). We shall now find the argument of this exponential function.
Since the brightness is proportional to p^2 [Eqs. (2.25) and (2.52)], it is sufficient to find the value
of the logarithmic derivative of p^2 with respect to Φ. We shall assume that

---

*These features apply also to the nonradiative recombination centers.

$$\frac{d \ln p^2}{d \ln \Phi} \equiv 2 \frac{\Phi}{p} \frac{dp}{d\Phi} = 2r = \text{const.} \tag{2.58}$$

Then, obviously,

$$p \propto \Phi^r = \exp\left(r \frac{eU}{kT}\right). \tag{2.59}$$

This means that if we approximate the dependence R(U) by an exponential function, the argument of this function near the maximum of the quantum efficiency will differ from eU/kT by a factor of r.

We shall find the derivative, utilizing the fact that the condition (2.19) is obeyed near the maximum of the quantum efficiency. We shall also make use of Eqs. (2.45) and (2.6) and allow for the inequality $\gamma_l \ll 1$. Differentiating Eq. (2.6) with respect to Φ, we obtain,

$$\frac{\Phi}{p} \frac{dp}{d\Phi} = 1 - \frac{\Phi}{n} \frac{dn}{d\Phi}. \tag{2.60}$$

The expression for $dn/d\Phi$ can be found from Eq. (2.45). However, before differentiating this equality, we shall transform it to the more convenient form

$$\gamma_l \Phi^2 + \Phi n (n + n_l) - n^3 N_l = 0. \tag{2.61}$$

Differentiating with respect to Φ and substituting the expression obtained into Eq. (2.60), we find that

$$\frac{\Phi}{p} \frac{dp}{d\Phi} = 1 - \frac{\gamma_l \Phi}{n^2} \frac{13n + n_l}{3\gamma_l N_l - n_l}. \tag{2.62}$$

Expressing Φ in terms of n by means of Eq. (2.19) and substituting the expression obtained into Eq. (2.61), we obtain an equation for n which has only one nonvanishing root

$$n \big|_{n_{\max}} = \frac{\gamma_l N_l - n_l}{2}. \tag{2.63}$$

Since n > 0 in all cases, the conditions for the maximum efficiency can be achieved only if

$$\frac{n_l}{N_l} < \gamma_l. \tag{2.64}$$

i.e., if the luminescence centers are sufficiently deep. Substituting Eq. (2.63) into Eq. (2.62) and replacing $\gamma_l \Phi/n^2$ with unity in the latter equation, we find that

$$\frac{\Phi}{p} \frac{dp}{d\Phi} = \frac{1}{4} + \frac{n_l}{4\gamma_l N_l}. \tag{2.65}$$

Finally, using Eq. (2.6), we find that when the excitation level is such that the efficiency is close to its maximum value, the brightness depends on the voltage in accordance with the law

$$R \propto \exp\left[\frac{eU}{kT}\left(\frac{1}{2} + \frac{n_l}{2\gamma_l N_l}\right)\right]. \tag{2.66}$$

It follows from the condition (2.64) that the voltage dependence of the brightness near the maximum of the quantum efficiency of electroluminescence can be approximated by an exponential function whose argument ranges from eU/kT to eU/2kT. When the depth of the luminescence centers is increased, the efficiency maximum shifts in the direction of the less steep part of the brightness-voltage characteristic.

The voltage dependence of the brightness described by Eq. (2.66) would be unsuitable for practical injection light sources. This is because the brightness depends very strongly on the applied voltage and on the temperature. The brightness changes by a factor of 2 when the voltage fluctuates by 1% or the temperature alters by 2-3 deg C. This inconvenient property can be minimized by reducing the quantum efficiency. The maximum efficiency lies somewhere near the point a_2 of the brightness−voltage characteristic (Fig. 13). At this point, the brightness-voltage characteristic can be approximated by an exponential function in a very narrow range of voltages, and at higher voltages the characteristic is horizontal. Therefore, if we increase the voltage so as to go over to the horizontal region, the dependence of the brightness on the voltage and temperature[*] ceases to be very strong (even if the horizontal region is not as prominent as in Fig. 13). If the semiconductor contains few undesirable impurities (for example, 10^{15} cm^{-3} or even 10^{14} cm^{-3}), such an increase in the voltage does not result in an appreciable reduction of the efficiency although the number of nonradiative recombination events will increase very considerably.[†]

We shall now consider recombination in the space-charge region near the boundary between the π-and n-type regions and we shall determine the maximum permissible concentration of deep quenching donors in this region.

The voltage drop in the space-charge region is

$$\psi_{\pi n} = kT \ln \frac{n_n}{n_\pi},$$ (2.67)

where n_n and n_π are the electron densities in the n- and π-type regions.

The width of the space-charge region can be calculated by solving Poisson's equations under the same assumptions that are made in the case of an ordinary p−n junction (see pp. 153 and 154 in [13]). Then,

$$d_{\pi n} = \sqrt{\frac{\varepsilon \psi_{\pi n}}{e^2} \frac{n_n + N_{\mathrm{as}}}{n_n N_{\mathrm{as}}}},$$ (2.68)

where ε is the permittivity and e is the electronic charge.

We have assumed that the negative space charge on the π-type side is concentrated entirely at shallow acceptors and that the density of this charge is independent of the coordinate. The positive charge on the n-type side is assumed to be concentrated at shallow donors, which should be fully ionized. Within the thickness of the space-charge region $d_{\pi n}$, the ratio n/p ranges from γ_l to $4\gamma_l n^2/(\gamma_l N - n)^2$, whereas the product np is constant and equal to $(\gamma_l N_l - n_l)^2/4\gamma_l$ throughout the space-charge region. This variation in the ratio n/p may extend over 12 orders of magnitude, from 10^{-6} to 10^6, because n_n is of the same order of magnitude as N_l. In a layer where

$$\frac{n}{p} = \gamma_l,$$ (2.69)

---

[*] The voltage and temperature dependences are interrrelated because the brightness is governed by $\Phi \propto \exp(eU/kT)$.

[†] For example, if the maximum efficiency is 99%, an increase in the number of nonradiative recombination events by a factor of 10 will reduce the efficiency by only 9% (to 90%).

rapid recombination occurs at deep donors. The effective width of this layer corresponds to a potential drop by a value which is of the order of $\pm kT$. Assuming, for the sake of simplicity, that the electric field in the space–charge region is constant,[*] we shall find the effective width of the enhanced-recombination layer

$$\Delta x \approx \frac{2kTd_{\pi n}}{\psi_{\pi n}} = 2kT \sqrt{\frac{\varepsilon}{e^2 \psi_{\pi n}} \frac{n_n + N_{as}}{n_n N_{as}}} \approx 2 \cdot 10^{-7} \text{ cm } . \qquad (2.70)$$

The recombination flux via the deep donors, per cm^2 of the junction area, calculated using Eqs. (2.26) and (2.29), as well as

$$n_{\pi n} p_{\pi n} = np, \qquad (2.71)$$

is given by

$$I_{\pi n} = N_d \sqrt{A_d C_d np} \, \Delta x. \qquad (2.72)$$

It follows from the inequality (2.4) that this flux should be much less than the recombination flux in the π-type region, whose value, calculated per cm^2 of the area, can be found from

$$J_\pi = I_l l_\pi, \qquad (2.73)$$

where l_π is the thickness of the π-type region. The Joule losses in the π-type region will be small if the thickness of this region is approximately equal to the diffusion length of the deficient carriers $L_{n\pi}$:

$$l_\pi \approx L_{n\pi} = \sqrt{\frac{D_{n\pi} n}{I_l}}, \qquad (2.74)$$

where $D_{n\pi}$ is the diffusion coefficient of electrons in the excited π-type region.

Substituting Eqs. (2.74) and (2.34) into Eq. (2.73) and using Eq. (2.29), we find

$$J_\pi = \sqrt{\frac{1}{2} D_{n\pi} N_l C_l np}. \qquad (2.75)$$

We shall substitute the values obtained in this way into Eq. (2.4), and we shall impose the following restriction on the maximum concentration of the deep donors in the space–charge region of the π-n junction:

$$N_d \ll \frac{1}{\Delta x} \sqrt{\frac{D_{n\pi} N_l C_l}{A_d C_d}} \approx 3 \cdot 10^6 \sqrt{N_l} \qquad (2.76)$$

Here, N_d and N_l are given in cm^{-3}. It follows that if the concentration of the luminescence centers is 1×10^{18} cm^{-3}, a semiconductor must be purified to ensure that the concentration of quenching donors is less than 3×10^{16} cm^{-3} if an efficiency of 90% is desired. This concentration is close to the maximum permissible value of the concentration of quenching centers in the π-type region.

In that part of the space–charge region where $n/p \approx \gamma_a$, rapid recombination occurs at the

---

[*]A more detailed analysis of this simplified approach is given on pp. 182–193 in [13]. A solution obtained without this simplifying assumption can be found in [14].

deep acceptors. Therefore, the restriction imposed on the maximum concentration of deep acceptors is the same as that for donors.

The electron density in the space-charge region of the p–π junction is several orders of magnitude lower than the hole density. Therefore, the restrictions that have to be imposed on the maximum concentration of the quenching centers are much less rigorous than those in the space-charge region of the π–n junction.

We shall now find the maximum permissible concentrations of deep impurities in the n- and p-type regions. In these regions, the electric field is relatively weak and it can be ignored in the equations of continuity. The solutions of these equations for the n- and p-type regions, in which the minority-carrier current is governed by diffusion and recombination processes, are identical with the solutions for a p–i–n junction [8]. The recombination current in the n-type region can be described by the standard formula if the minority-carrier lifetime is independent of the excitation level and is governed by deep acceptors:

$$J_n = \sqrt{D_{pn} N_a C_a \frac{np}{n_n}}, \qquad (2.77)$$

where D_{pn} is the diffusion coefficient of the holes in the excited n-type region. A similar expression can be derived for the recombination of electrons (injected into the p-type region) at deep acceptors.

It is evident from Eq. (2.77) that the recombination currents in the n- and p-type regions will be small if these regions are heavily doped with shallow donors and acceptors and if they contain the minimum possible number of other (deep) impurities. When the doping level is increased, the majority-carrier density increases and the diffusion coefficient decreases. These two effects reduce the recombination current. A reduction of the concentration of the deep impurities increases the minority-carrier lifetime and this also reduces the recombination current. Even if a semiconductor is so pure that its minority-carrier lifetime is inversely proportional to the dopant concentration, an increase in this concentration would still be desirable because Eqs. (2.3) and (2.4) contain the square root of the lifetime whereas the free carrier density is present as the first power. This applies, to a decreasing degree, right up to the degeneracy threshold.

We shall now substitute Eqs. (2.75) and (2.77) into the inequality (2.5). Applying Eq. (2.33) and making certain transformations, we obtain

$$N_a \ll \frac{1}{2} \frac{D_{n\pi}}{D_{pn} C_a} \,^1\!l \left(\frac{n_n}{p}\right)^2 N_l. \qquad (2.78)$$

We shall ignore the difference between the diffusion coefficients, and we shall make the inequality more stringent by assuming that $n_n \approx p$. Since $A_l \approx C_a$ [Eq. (2.22)], we obtain

$$N_a \ll \frac{1}{2} N_l. \qquad (2.79)$$

We see that the restrictions imposed on the concentrations of the deep impurity centers in the p-type region (and similarly in the n-type region) are the least rigorous. Such deep centers do the greatest harm in the space-charge region of the π–n junction.* If deep donors and acceptors are distributed uniformly across a sample, their concentration must not exceed 10^{16} cm^{-3} for a quantum efficiency of 0.9 or $\sim 10^{15}$ cm^{-3} for an efficiency of 0.99.

---

*If the luminescence centers are deep acceptors, this reasoning applies to the space-charge region of the p–π junction.

We shall now estimate the output power density (irradiance) of an injection source of light. We shall assume that

$$p \approx N_l \approx 3 \cdot 10^{18}\,\text{cm}^{-3},\ \frac{n}{p} \approx \gamma_l \approx 10^{-6}. \tag{2.80}$$

The above expressions represent favorable conditions for the luminescence centers. The free-carrier mobility in phosphors is usually of the order of 100 $\text{cm}^2 \cdot \text{V}^{-1} \cdot \text{sec}^{-1}$ and, therefore, the room-temperature diffusion coefficient is ~ 3 cm^2/sec. We shall use these estimates to determine the output power density

$$B_l = h\nu_0 J_\pi \approx 1\ \text{W/cm}^2, \tag{2.81}$$

where $h\nu_0 \approx 2$ eV is the average energy of the emitted quanta.

We can see that the output power density of a source of moderate size is fairly high. It is sufficient to have a phosphor area of 3–4 cm^2 in order to produce the same luminous flux as an incandescent lamp consuming 100 W. However, these parameters do not represent the maximum that can be achieved. There are some luminescence activators with capture cross sections up to 10^{-18} cm^2. The use of such activators should make it possible to increase the output power density by two orders of magnitude compared with the values just given. If the applied voltage is less than the energy of the emitted quanta, a light source of this type will cool because of the negative Peltier effect and its efficiency will be higher than 100%. However, under these conditions, the output power density will decrease because it is an exponential function of the voltage. Therefore, the maximum radiant efficiency of injection electroluminescence is achieved at an output power density lower than that corresponding to the maximum quantum efficiency.

The luminescence spectrum of an injection source can be made to fit any specification. All that is necessary is to introduce several activators which give rise to the required spectral bands. However, since the parameters of different luminescence centers are, naturally, different, the luminescence spectrum of a crystal containing many activators will depend strongly on the voltage. Therefore, it is more convenient to produce a complex spectrum by connecting in series several different sources emitting different colors. For example, if it is desired to have a white spectrum without any deep troughs, it is necessary to connect in series four or five different injection sources.

We shall now formulate briefly the conclusions reached on high-efficiency sources of light based on the injection electroluminescence mechanism.*

1. Injection light sources should be made of wide-gap (2–3 eV) semiconductors containing activators which can emit luminescence in the visible range. The quantum efficiency of the luminescence should be high.

2. The semiconductor should be used to make a $p-\pi-n$ junction in which the π-type region is a phosphor and the p- and n-type regions have a low resistivity. The thickness of the π-type region should be of the order of the diffusion length of carriers in this region.

3. The activator and coactivator concentrations in the phosphor should be of the order of $10^{18} - 10^{19}$ cm^{-3} (higher concentrations result in quenching) and the concentrations of harmful quenching impurities should not exceed 10^{16} cm^{-3}.

4. Several sources of this type, having different spectra, can be used to build a lamp emitting white light. A lamp of this kind should have a radiant efficiency close to unity.

---

*We shall not consider the case when luminescence centers in a phosphor are donor−acceptor pairs. The light sources made of such phosphors have characteristics similar to those discussed in the present paper [15].

Conclusions

We have considered two ways of improving light sources. The efficiencies of the improved sources should be close to the limits set by thermodynamics.

We shall now summarize briefly the expected properties of the optimal light sources of both types (thermal and luminescent) considered in the present paper.

TABLE 1

Property	Semiconductor incandescent lamp	Semiconductor injection lamp
Luminous efficiency	50-60 lm/W (50-70 lm/W with a vacuum jacket)	200-220 lm/W (for white light)
Voltage dependence of luminuous efficiency	Has a maximum	Has a plateau
Color	Same as tungsten lamp	Any color (white light can be obtained using several sources)
Voltage dependence of color	Becomes whiter with increasing voltage	Independent of voltage
Change in brightness for a change in voltage by 5%	By 35-50%	By a factor of 3-9 (can be stabilized by a series choke)
Change in brightness for a change in current by 5%	By 35-50%	By 5%
Current−voltage characteristic	Nonlinear and symmetrical	Same as diode
Flicker at alternating voltage	Same as tungsten lamp	Emits only during one half-period (flicker reduced by parallel connection of second lamp)
Service life	Of the same order as tungsten lamp	Unlimited

There is little doubt that both types of lamp will find wide applications. All that remains is to produce them.

Literature Cited

1. M. A. Weinstein, J. Opt. Soc. Amer., 50:597 (1960).
2. M. V. Fok, Opt. Spektrosk., 13:612 (1963).
3. E. Kauer, Philips Tech. Rev., 26:33 (1965).
4. Yu. N. Nikolaev, ZhETF Pis. Red., 4:474 (1966).
5. V. S. Vavilov, Effects of Radiation of Semiconductors, Consultants Bureau, New York (1965).
6. É. E. Violin and G.F. Kholuyanov, Fiz. Tverd. Tela, 6:1696 (1964).
7. W. Shockley and W.T. Read Jr., Phys. Rev., 87:835 (1952).
8. A. Herlet and E. Spenke, Z. Angew. Phys., 7:195 (1955).
9. V. V. Antonov-Romanovskii, Kinetics of Photoluminescence of Crystal Phosphors [in Russian], Nauka, Moscow (1966).
10. M. K. Sheinkman, I. Ya. Gorodetskii, and I. B. Ermolovich, Fiz. Tverd. Tela, 7:3134 (1965).
11. R. A. Logan, M. Gershenzon, F. A. Trumbore, and H. G. White, Appl. Phys. Lett., 6:113 (1965).
12. É. E. Violin and G. F. Kholuyanov, Fiz. Tverd. Tela, 6:593 (1964).
13. G. E. Pikus, Fundamentals of the Theory of Semiconductor Devices [in Russian], Nauka, Moscow (1965).
14. C. T. Sah, R. N. Noyce, and W. Shockley, Proc. IRE, 45:1228 (1957).
15. K. Maeda, J. Phys. Chem. Solids, 26:595 (1965).